Studies in Emotion and Social Interaction

Paul Ekman
University of California, San Francisco

Klaus R. Scherer
Justus-Liebig-Universität, Giessen

General Editors

The Meaning of Primate Signals

Studies in Emotion and Social Interaction

This series is jointly published by the Cambridge University Press and the Editions de la Maison des Sciences de l'Homme as part of the joint publishing agreement established in 1977 between the Fondation de la Maison des Sciences de l'Homme and the Syndics of the Cambridge University Press.

Cette collection est publiée en co-édition par Cambridge University Press et les Editions de la Maison des Sciences de l'Homme. Elle s'intègre dans le programme de co-édition établi en 1977 par la Fondation de la Maison des Sciences de l'Homme et les Syndics de Cambridge University Press.

The Meaning of Primate Signals

Edited by Rom Harré *and* Vernon Reynolds

University of Oxford

The right of the University of Cambridge to print and sell all manner of books was granted by Henry VIII in 1534. The University has printed and published continuously since 1584.

Cambridge University Press

Cambridge
London New York New Rochelle
Melbourne Sydney

Editions de la Maison des Sciences de l'Homme

Paris

Published by the Press Syndicate of the University of Cambridge
The Pitt Building, Trumpington Street, Cambridge CB2 1RP
32 East 57th Street, New York, NY 10022, USA
296 Beaconsfield Parade, Middle Park, Melbourne 3206, Australia
and
Editions de la Maison des Sciences de l'Homme
54 Boulevard Raspail, 75270 Paris Cedex 06

First published 1984

Printed in Great Britain at the University Press, Cambridge

Library of Congress catalogue card number: 83–23219

British Library Cataloguing in Publication Data
The Meaning of primate signals – (Studies
in emotion and social interaction; no. 4)
1. Primates 2. Animal communication
I. Harré, Rom II. Reynolds, Vernon
599.8'0459 QL737.P9
ISBN 0 521 25944 4 hard covers
ISBN (France only) 2 7351 0072 3

WV

Contents

Contributors

Edwin Ardener
Institute of Social Anthropology
Oxford University

Pamela J. Asquith
Primate Research Center
Kyoto University

Michael Chance
Sub-department of Ethology
University of Birmingham

Dorothy L. Cheney
Department of Anthropology
University of California, Los Angeles

Frans B. M. de Waal
Wisconsin Regional Primate Research Center
Madison, Wisconsin

G. Ettlinger
Fakultät für Psychologie und Sportwissenschaft
Abteilung Psychologie
Universität Bielefeld

Rom Harré
Sub-Faculty of Philosophy
University of Oxford

Roy Harris
Worcester College
University of Oxford

Eric Jones
Sub-department of Ethology
University of Birmingham

Justin Leiber
Department of Philosophy
University of Houston

M. Maurus
Max-Planck-Institut für Psychiatrie
München

D. Ploog
Max-Planck-Institut für Psychiatrie
München

Duane Quiatt
Department of Anthropology
University of Colorado

Vernon Reynolds
Department of Biological Anthropology
University of Oxford

Robert M. Seyfarth
Department of Anthropology
University of California, Los Angeles

H. S. Terrace
Department of Psychology
Columbia University

Preface

The papers in this volume reflect the growing interest throughout many academic specialities in the study of animal communication. The original impetus for a joint approach among philosophers, linguists and ethologists came from the meeting convened at the Werner-Reimers-Stiftung in 1977 to explore the possibilities for human ethology. That meeting issued in the publication of M. von Cranach, K. Foppa, W. Lepenies & D. Ploog, *Human Ethology: Claims and Limits of a New Discipline*, Cambridge University Press, 1979.

The work reflected in this interdisciplinary volume is the result of an intensive discussion amongst the various authors, which took place in Bad Homburg in January 1981. It was made possible by the generous support of the Werner-Reimers-Stiftung. That meeting was planned by Professor J. Aschoff, Professor D. Ploog, Dr. M. Maurus, Dr. V. Reynolds and Mr. R. Harré. Though this volume is not to be thought of as the proceedings of a conference it could not have been produced without the close interaction between the authors.

Besides the participants who have contributed chapters to this volume, a number of scholars were present at the conference and have contributed their comments, namely Professor H. Kummer, Professor K. Scherer, Dr. F. de Waal, Professor G. Ettlinger and Dr. W. Mason.

The word 'primate' as used in the title refers to all primates (including humans). Individual authors vary in their usage, sometimes referring to non-human primates only when they use the word 'primate'. We hope that the sense in which the word is used is in each case clear from its context.

We should finally like to thank all members of the Werner-Reimers-Stiftung, in particular its director Dr. K. von Krosigk and its secretary Mrs. G. Söntgen for their continuous support.

V. REYNOLDS

Oxford, 1983 R. HARRÉ

Introduction

The title of this book, like that of the conference from which it arose, expresses a very real problem for students of primate behaviour and human action. In a nutshell, it is a problem that arises from our humanness. Each of us has personal experience (and therefore knowledge) of the complex cognitive thought processes that go on in his or her head. We know we think, and we know our thoughts to be long, convoluted and interminable. We also know that these thoughts underlie our actions. They do not perhaps entirely determine our actions but they do guide them. Further, without saying that we always think in language, it is clear to us from personal experience that much of our thinking is word-thinking.

When we study the lives of animals, we see each species behaving in its own particular ways. Ethology, a branch of zoology primarily concerned with behaviour, has given us, over the last fifty years, an explanatory framework for the understanding of animal communication. In evolutionary terms we can look for the origins of signals by searching for their selective advantages. In ontogenetic terms we can study how the behaviours of adults develop, through a combination of innate and learned elements, during pre-adult life. We can discover the external causes of the things animals do, and the effects of their behaviours on others. But ethology has not, because of its zoological origins, given us an adequate psychology of animal behaviour.

The core of any psychology is the workings of the mind. We have a unique entrée into the human mind: our personal subjective experience. From this starting point, which we must accept as a valid source of data (for if our own experience is not valid then what is?), we can attempt to generalise outward into the minds of others. If we could not do this we could not explain the actions of other people. What they do makes sense only in terms of how it compares with what we would do in a comparable situation. That is the basis of common-sense, or 'folk' psychology.

But in fact, as we also know, what other people do is not always explicable in terms of what we ourselves would or would not do. Anthropology has given us glimpses into psychologies other than our own. In such cases, it is to the social world, the world of received ideas, of given knowledge, of prescriptions and proscriptions, totems and taboos that we have to turn to find the sources of ideas and actions. And, by reflecting on ourselves again, we know that in our case, too, the structure of our aims and ambitions, our loves and hates, has its origins not within ourselves but in the social world.

Thus we can make progress in an analysis of ourselves; we can formulate a human psychology in which cognitive and motivational elements are fused into coherent chains of thought and action. In short we can to some extent come to know ourselves. But how is it to be with our closest relatives in the animal kingdom, the non-human primates? Can we come to known their psychology, and if so, how? One thing is certain, and has been emphasised by ethology throughout its history: that we cannot allow ourselves the luxury of an unthinking anthropomorphism. There is no *a priori* reason why animals should process the information reaching them, digest it, and respond to it, in the same ways we do. Certainly we would not expect to find human-like thoughts in an insect, fish or bird. It is more difficult to be so sure in the case of mammals, especially non-human primates. Why not make the *a priori* assumption that they do think in human-like ways?

The whole of science is against such an assumption. Only after the use of Occam's razor could such a conclusion be reached, i.e. by eliminating all simpler explanations. That, at least, is the standpoint of Western science.

Interestingly, the Japanese primatologists of the 1950s, who pioneered the detailed study of primate social life, were anthropologists, not zoologists. Imanishi, Itani, Kawai, all broke into the world of the Japanese macaques without fear of anthropomorphism, without the scruples of Western zoology. In so doing they discovered a creature remarkably like ourselves, in its preoccupation with rank and its close bonds of kinship. Monkey matrilineages were known to the Japanese and were described in English some twenty years before Western science had advanced to the point where such concepts were acceptable. Asquith's contribution (this volume) pays attention to the case of Japanese primatology and helps explain how monkey kinship came to be first discovered in Japan. Western science was only able to incorporate kin behaviour when it had advanced theoretically to the necessary point, and this came about through advances in genetical theory and the development of the idea of inclusive fitness. There remain today reservations among Japanese scholars about the use of genetical theory to explain monkey kinship: their theories are more psychological, involving concepts such as 'identification'. Such processes still await the stamp of approval from mainstream Western science. The reason is as it was earlier: no clear mechanism has yet been discovered, and without a mechanism we have speculation rather than science.

If science is stubborn, scientists in most cases are adventurous. They try to keep an open mind, they try to think of situations and experiments to test the hypotheses that suggest themselves. For many years now, primatologists have made close studies of primates living in social groups and have discovered the extraordinary complexity of their communication systems, their relationships, and the structures of their societies. A theory based on automatisms cannot hope to account for the subtleties of primate behaviour in a social setting. There is constant monitoring of each individual by the others. There are careful choices of who to interact with, where to sit, when to take food and when to hold back. Though they may not be aware of social rank as such, monkeys are aware of who dominates them and who is dominated by them. They know how to time a sexual advance. When they deceive, they do not, as Quiatt shows, act out programmed innate patterns of evolved deception (in contrast to the lapwing with its 'broken-wing display'). They practise carefully executed acts of deceit: pretending, for instance, not to know where a piece of hidden food is until all competitors are out of sight and then grabbing it.

Though lacking language, they do not lack thought. But do they possess something which, though not yet language in the proper sense, is nevertheless something more than the kind of signalling system found throughout the animal world and so intensively studied by ethologists? Does the nature of primate communication converge in any respects with human linguistic systems? Hitherto, such convergence has been seen at the level of what in humans is called 'non-verbal communication'. This focus has masked a very different area of convergence that emerges from studies like those of Cheney and Seyfarth on wild vervet monkeys. These animals are able to communicate rather precise information by means of inflexions in their calls. They can convey information about the type of referent and about the location of the referent. In so doing, they appear to be starting on a category system. As Harris points out, they make subdivisions of existing categories, and these they apparently learn during development. Such categorisation is very like one quality of human languages.

Certainly, in captivity, primates can categorise. The studies of chimpanzees who are taught American Sign Language (ASL) or other sign languages show that a one-to-one matching between objects and signs or symbols is quite manageable. Just what the matching consists of and how it compares with human use of words to designate objects and abstractions is a topic discussed by Terrace, and it is vital for the progress of our understanding of primate thought that we should know what it is that they are doing when they use signs and symbols.

The richest use of such mental powers as non-human primates possess will always be in their own social lives. A 'talking' chimp is a deprived chimp, deprived of its own kind and the richness of its own way of life. The social life of monkeys in a captive colony is a rich field of information flows, in which each member is caught in a particular situation by the forces of the actions of the other members of the group. As Reynolds describes, there is an overt pattern of signals from individual to individual, but beneath there are long-term dramas taking place, in which individuals are locked in struggle with each other. To explain these day-to-day co-ordinated patterns of behaviour as automatisms seems to be stretching credibility.

But if we are to impute a world of meaning to non-human primates, in which they know what is going on around them in some detail, and are aware not only of their own motives but of the motives of others around them, how are we to describe what they do without humanising them completely? The use of our own, human-based, language will inevitably humanise them to some extent, and this, as Asquith shows, cannot be avoided. Knowing this, should we resign ourselves to language use? Harris states that we have no alternative. If that is so, we are going to need to be very precise indeed in our use of language in order to convey exactly what primates are capable of understanding, what factors they can take into account, and how rich their signals are in underlying cognitive meaning.

Part I
The setting of the problem

Introduction

It has long been scientific dogma that we cannot know how animals think, so much so that animal psychology has almost entirely abandoned the issue. Stimuli of known characteristics are applied to experimental animals, in a known environment, and the responses are measured. What goes on in the animal's head may be measured in the form of EEG impulses, but what goes on in its mind is considered unknowable.

Ethology has added a vast richness to this experimental paradigm. We do not need to isolate an animal or to control the input side; we can, by patient observation in natural conditions, discover that certain environmental features are attended to by an animal, and that certain predictable responses follow. In general, the relation between the stimulus and the response is adaptive: it is conducive either to the survival or to the reproductive success of the animal. So we can assume that in the head of the animal there are neural programmes that assess environmental features and selectively activate responses appropriate to them. If we were to talk of mind, thoughts or ideas, we would need to describe them using a functional terminology: the animal's mind is full of useful solutions, and there are three strong urges in its composition – the urge to escape danger, the urge to find food, and the urge to mate.

It is possible, however, to draw an analytical dividing line in the animal mind, between those elements that are relatively automatic, inflexible products of genetic programming, and those that are flexible, situation-specific products of cognitive assessment. This emerges clearly from Quiatt's chapter on the subject of 'deception'. The deception practised by a stick insect pretending to be a stick is of the former sort: it is a direct product of natural selection. The deception practised by a zoo chimpanzee which keeps its mouth full of water until a human is within range and then spits it at him is not a product of natural selection but a well-planned cognitive action. They are both strategies, but in very different senses: the former is a genetic strategy, the latter an individual one.

Quiatt also discusses the question of intentionality. It would be quite wrong to say the stick insect intends to look like a stick; but is it wrong to say the chimpanzee intends to douse the unsuspecting onlooker? At first sight it seems the most parsimonious explanation, but we must ask ourselves whether the whole process might not happen in just the same way without an intention at all. The problem is that we have no direct evidence of the intention itself. And we don't even really know what an intention is. Is it a discrete entity that exists at a moment prior to some action? Or is it a substrate accompanying the action itself? In general, we have to infer the existence of an intention from a close analysis of the context and the actions of the individuals concerned. If, time after time, we infer that an intention probably did precede or underlie the actions we see, then we should not dismiss this inference because of the lack of proof. We should investigate it. Whether primate intentionality is the same as human intentionality is another matter, but there seems no reason to deny the possibility of some degree of overlap between primate intentions and some kinds of human intentions.

When we study primate social behaviour in the wild, there is less evidence of treachery and deceit than is seen among captive animals, probably because there is a lower level of stress in the group (which is less crowded and has a far richer environment). What can we glean from the wild about the meanings underlying primate signals? Seyfarth and Cheney have both studied vervet monkeys in the wild in Africa, with a mind to this problem. Seyfarth describes how differences of meaning can be discovered by use of an ingenious field-experimental technique involving the play-back of vocalisations from a hidden loudspeaker to monkeys in the wild. Using the well-established principle of judging the meaning of a signal by the effects it has on others, he was able to show that there are surprisingly specific shades of meaning in primate calls, to the point where comparison with human language seems appropriate. Seyfarth also found that the meaning of the calls resides in the acoustic qualities of the calls themselves and not in features of the context.

Cheney takes up the matter of primate 'classification' of aspects of the environment. Similar-sounding acoustic units can refer to quite different objects in the environment, and different-sounding units can designate similar referents. These findings indicate a degree of comparability between monkey calls and human language not hitherto suspected, and they give us a glimpse into the nature of the monkey mind, the distinctions it holds. Further, there is evidence of the refinement of these distinctions during ontogeny. The young vervet gives a generalised alarm call to a variety of birds, later coming to restrict them to predatory birds only; Cheney suggests that the participation of the mother in cases where there really is a predator may be important in refining the call. We know that in humans, too, the broad categories of childhood are refined by learning and experience into the finer discrimination of adulthood. Besides classifying their environment, monkeys also distinguish each other, and classify others into certain categories, i.e. offspring v. other infants, or group-member v. outsider. All in all they appear to have a far more complex apperception of the world than we had hitherto suspected and, importantly, we are beginning to find ways of probing into it.

The question now arises as to how the results of these researches should be described. How we choose to describe them will determine in part how we explain them; and in its turn, how we explain them will define the outline strategy for further research. In choosing to describe primate actions as intentional we open up the possibility of explaining what some primates do as the realisation of intentions in action. In so doing we define an outline research strategy – find the intention. In this context 'intention' is a theoretical concept and we want to know whether it has some kind of real referent, in the appropriate domain. In the natural sciences a more or less standard research strategy is now invoked. One asks in what respects a new theoretical concept, say 'natural selection', is similar to or different from the concept 'domestic selection', from which it was derived. To find whether natural selection really happens, the status of the process (if it exists) has to be worked out, otherwise we scarcely know were to look. If it is conceived as *very* like domestic selection, we might look for it in the alleged pronouncements of the Great Breeder in the Sky. If it is rather different from domestic selection we may look in nature for selection mechanisms. Research strategy is determined by conceptual structure. In the enterprise described in this book the authors are trying to provide a clear enough answer to the question of how similar (and how dissimilar) animal intentions are to those of humans to permit the formulation of an accurately directed research strategy.

We know our intentions in the prerepresentations (both iconic and conceptual) by which we define our goals. We know the intentions of others in their declarations, and these need not always be verbal. In many human contexts public declarations of intention are psychologically more important than the private representations of goals. Do animals declare (and conceal) their intentions? The answer seems to be that they do. But when they are so concealing them (that is privately representing but not publicly displaying) how is the representation achieved? Iconically in the form of concrete images (say of wet and bedraggled zoo visitors, shortly to be drenched by the water concealed in the mouth), or conceptually? Do animals not only have inten*t*ions but inten*s*ions, conceptual ways of representing their goals?

Many theorists of psychology would hold that there is a strict relation between the capacity to use concepts for thought and the acquisition of language, the latter conceived as an abstract and arbitrary system of signs. Part of the problem area the authors of this volume are addressing is the exact shading of the language-using/non-language-using distinction as it is applied on the human/primate boundary. In this context the issue of whether 'signing' chimpanzees 'have language' assumes great importance. If they have, there is at least a necessary condition satisfied for their having conceptual (intensional) thoughtways. It seems that the question of whether some non-human primates 'have language' does not admit of a simple or clear cut answer. It looks now as though even the smartest 'signing' chimpanzee lacks syntactical capacity, but does have semantical capacity of quite a high order. Does this show that such animals are capable of deploying (or, more strongly, do actually employ) an abstract and arbitrary system of signs in representing that which they plan to bring about? We hope to give at least a partial answer to that question in this volume.

A. The problem of animal mentation

1. Devious intentions of monkeys and apes?

DUANE QUIATT

> It is commonly stated that no animal can use its communication system to tell a lie. Of course, a lie requires intention to deceive, so that to judge whether variability in animal communication behavior is 'noise' or prevarication requires knowledge of the animal's intention. (Griffin 1976:45)

Griffin, in the work from which the above quotation is taken, argues that animals must have intentions, 'mental images of future events in which the intender pictures himself as a participant and makes a choice as to which image he will try to bring to reality' (1976:5). Griffin notes a circularity in arguments which rule out intention *a priori* and then take its absence 'as evidence that animal communication is fundamentally different from human language' (1976:5). With greater confidence than many biologists are willing to place in an evolutionary continuity that extends to the nature and internal experience of individual organisms, Griffin argues strongly for a fundamental likeness of cognitive aspects of animal communication, whatever animals are doing the communicating.

Intention is usually assumed to underlie deception in human communicative behavior. Knapp & Comadena, in a recent review of theory and research on deception in human communication, observe that '*Many of the studies on lying and deception seem to operationalize the act as: the conscious alteration of information a person believes to be true in order to significantly change another's perceptions from what the deceiver thought they would be without the alteration*' (1979:271, authors' emphasis). Complex mental processes are taken for granted in the studies reviewed by Knapp & Comadena, and they are not of the sort that lend themselves to physical description. Measurement of physiological responses during cognitive functioning has a long and respectable history (McGuigan 1978, 1979), but the central processes of cognition remain inaccessible to measuring instruments. The increasing acceptance of cognitive terms and concepts in discussions of human and nonhuman animal behavior (McCall & Simmons 1978; Hulse, Fowler & Honig 1978) owes more to the logical compatibility of computer models of symbolic representational systems with theories of cognitive structure than it does to progress in the observation and description of behavior. The functional nature of definitions of cognition, based on observations of behavior, limits the meaning of such concepts to the range of observations upon which they are based. 'A positive description of cognitive mechanisms or processes common to the defining observations still seems to be lacking . . . [and probably] will not be provided by further, more detailed, or even more insightful description of experimental procedures or paradigms. The distinctive characteristics of cognitive processes will be more directly established only by the mechanisms which are observed or *assumed* to explain the behavior – in short, by theory' (Honig 1978, my emphasis). In short, the problem remains that of how to represent the mind's interior so that we can be clear about and agree on what it is we think we mean when we speak of cognitive processes such as intention.

1.1. Reflective thinking and introspective accounts of mental states

Introspection, our own experiences and detailed accounts by others of remembering, forgetting, making decisions, and so on, give us reasonably direct access to the kind of mental images that we may associate with intention, and what this evidence lacks in 'objectivity' seems more than made up for by vivid clarity of detail. In such representations as Einstein's account of laboriously translating visual and 'muscular' images into conventional language and symbols, or Sir Flinders Petrie's claim to use of a mental slide-rule for mind's-eye computation of sums (both cited by Ferguson 1977), there is a self-conscious distancing of reflective imagery from images processed that corroborates the conviction we may have that human intentions (at least) and related cognitive activities cannot be accounted for simply in terms of learning theory, not in terms of any self-contained and usefully economical theory of direct or indirect behavioral response to external stimuli. Yet even those who concede to other animals, in principle or in degree, a self-consciousness similar to our own, are likely to hesitate before accepting at face value accounts which purport to reveal that consciousness of self at *work* in other than *Homo sapiens* even within the order Primates:

> In a recent conversation Maureen asked Koko [a gorilla], 'Where do gorillas go when they die?'
> Koko replied, 'Comfortable hole bye'.
> Then she asked, 'When do gorillas die?'
> 'Trouble, old'.
>
> (Patterson 1980: 112)

And:

P: What does the word Ron mean to you?
[It has previously been explained that 'Ron is a dominant male father figure and disciplinarian in Koko's life'.]
K: Obnoxious . . . bite.
. . .
P: Who is onni-Ra?
K: Trouble devil.
P: What is andy-Ca?
K: Koko candy.
P: Tell me something funny.
K: Koko-love (and kisses Ron on the cheek).

(Patterson 1980:113)

I want to emphasize that I am concerned here with a problem that relates to cognitive processes – reflective thinking and intention – to which a concept of self is basic, not with either (*a*) the issue of gorilla or chimpanzee or other animal language *per se*, or (*b*) the issue of whether human beings can converse with gorillas or chimpanzees or other animals. While the notion of distinct econiches would appear to place enormous difficulties in the way of holding meaningful conversations across species boundaries (Wittgenstein's frequently quoted aphorisms seem precisely relevant in this connection), if we *can* talk to the animals, then it seems not unreasonable that Koko in the instances cited should converse in something like idiomatic American English (a matter of translator decisions) and should understand pig-latin ('catching on' to sign transformations). But I find it perplexing and awkward, almost embarrassing, that the self from which Koko establishes a removal in her ruminations on death and in her essay at humorous sarcasm (an entity which we are thereby enabled to view from her perspective), should be so little distinguishable from my own self and from the selves of people whom I know. Of course there are neurophysiological commonalities that extend beyond species boundaries, and from this we may assume that closely related species have cognitive structures or features of cognitive structures in common, so that as Gregg has put it, 'work on comparative aspects of nonhuman primate cognition is . . . a matter of the degree to which we wish to disagree' (Gregg 1971). All the same, and despite evident continuities of behavior between humans and other living primates, there may be great gaps about which, lacking intermediate populations to investigate, we can know nothing (Geschwind 1971). That is a most important point to keep in mind, especially if vocal language is a late development (Livingstone 1973), with gestural language being characteristic of chronospecies intervening between ancestral apes and *Homo sapiens* (Hewes 1973, 1977).

What disturbs me, then, about the dialogues with Koko is not that she and I may have thoughts in common – for it seems reasonable to suppose that animals who have so much else in common also think somewhat alike – but that there should be so little of the ape on her end of these conversations. This suggests either that a gorilla raised by humans is no gorilla or that our own limitations are such that even when we talk to gorillas we can neither ask nor discover what it is like to be one.

While it may be possible that 'At present, language allows perhaps the most direct channel for inter-species communication, and hence for a redefinition of categories. . . . [of] what is human and what is animal' (Patterson 1980:114), still, none of the evidence to date suggests that introspective accounts of reflective thinking and mental states are very useful for extending the study of cognitive processes across species lines.[1]

1.2. Deception in relation to intention

One way to talk about the meaning of a signal would be in terms of some goal entertained by the individual who transmitted it. However, it is difficult to relate signals to the intentions, if any, of nonhuman animal communicators, and ethologists have traditionally focused on responses as the crucial components of communication. According to Smith (1977), ethologists have occupied themselves 'almost solely with responses and their functional implications, with a consequent lack of research on the information contents of displays'. Such a charge may ignore the considerable attention paid by primate ethologists to the involvement of cognition in behavior of monkeys and apes, but little of that work has gone beyond attempts to demonstrate and categorize conceptual abilities (Gardner & Gardner 1971; Premack & Premack 1972; Rumbaugh, von Glasserfeld & Pisani 1973). June Leonard, who focused an experimental study of dominance behavior in 12 juvenile stumptail macaques on the strategies employed by individuals to achieve and maintain rank within the group, refers to that other sort of work as 'lip service' with no real 'marrying of cognitive concepts with social behaviour research' (1979:170).

But how are we to get at the intended meaning of signals, dealing with communicators who cannot speak to us of their intentions? In some circumstances it is possible to observe animals coercing companions toward the achievement of goals known to (or assumed by) human observers. Emil Menzel's oft-cited work with chimpanzees is very much to the point here. Menzel allowed individual chimpanzees to observe the placement of food and other items in a variety of circumstances, then released them, alone or with a group of naive companions, and noted how they used this information about their environment, including strategies for communicating and withholding it. One of Menzel's concerns in this work was cognitive mapping, which he treated as a structural concept logically independent of other sorts of concepts – developmental, functional, and evolutionary – all seen as essential to a complete psychology. 'Cognitive mapping', Menzel suggests, 'is . . . [not] a complete explanation of how animals are able to get around in the world. It is largely a metaphorical statement about what sorts of information they collect and how they organize it. . .' (Menzel 1978).

Menzel studied cognitive mapping mainly by observing how individual chimpanzees recovered items (or avoided feared items) the locations of which had earlier been revealed to them. How animals communicate such information, and under what circumstances they may withhold it, has been the main subject of the group studies. Smith has summarized succinctly these aspects of Menzel's work:

> The leaders appear to direct the others not with special display behavior, but primarily by the way in which they set out to go to a site: their pace, style of locomotion, and glancing patterns . . . intention becomes clearly evident in events when no other chimpanzee spontaneously follows the leader, especially if the leader is one of the smaller chimpanzees . . . unwilling to move out on their own. The leader will then glance from one to another of his companions, beckon with his hand or his head, walk backward toward the goal while orienting toward the group, and if all else fails try to pull or drag one of the others with him. *Intention in these cases can be assessed from the animals' attempts to manipulate each other's behavior in terms of their own known goals, the communicator continuously adjusting its performance to the kinds of responses it elicits.* (Smith 1977:263, my emphasis)

It is interesting that the behavior of leaders appears not to be characterized by any 'special display behavior' (though one might well argue that to pull or drag another chimpanzee toward a goal is rather special behavior) and that the main distinguishing feature of a leader's behavior is its seeming purposefulness (see also Chalmers 1980:81). Behavior which does involve special communicative displays and is similarly concentrated toward specific goals (as, for instance, in predator-defense, courtship, or grooming activity) also appears purposive to human observers presumably because it is unified, focused, and likely to elicit responses from within a narrow range. Such behavior can lead to 'surprising' outcomes, as when a male rhesus monkey is rebuffed in his attempt to mount a female who has just presented to him, or when a female with infant responds to grooming overtures and is victimized by a kidnapper (Quiatt 1979). Startled reaction by the signal-recipient to such a turn of events can suggest to a human observer that he or she was not alone in misconstruing the situation. *Present* behavior in rhesus monkeys turns out to be ambiguous; 'invitations' to groom may be formally indistinguishable from 'invitations' to mount (Quiatt 1966), and it is probably inevitable that they be misinterpreted from time to time.

Misinterpretation of a signal of course should not imply that the signal reflects behavior intended to deceive, or even behavior that was consciously intended. However, in the case of grooming by a prospective kidnapper, the grooming itself is likely to appear to a human observer perfunctory, no more proffered than is necessary to 'lull' the mother and afford access to the desired infant. Whether, in any such given instance, the groomer has in mind from the start detaching the infant from its mother or is simply attracted to it and ends up detaching it would seem to be impossible to determine, and at the same time would be critical to the issue of intention. In either case, the ultimate goal of the kidnapper is masked from the mother by grooming activity, and from her perspective it is a nice question whether she has been deceived or is simply confused. Menzel's work shows clearly that chimpanzees at least communicate a good deal more than information about their immediate emotional states, and it seems not unreasonable to ask under what circumstances information may be communicated that masks emotional states or that suggests an emotion or an intensity of emotion different from that which the communicator is actually experiencing.

Smith (1977:264) notes what he calls a 'disturbing possibility . . . that nonhuman animals might sometimes use displays to manipulate their recipients, intentionally misleading them'. If they do, Smith suggests, 'it is at least possible that ethological observers will detect it through contradictory patterns in the behavior that correlates with displaying . . . [though] ethologists have not yet caught nonhuman animals in intentional acts of lying with display behavior'. Of course ultimately we run up against the same problem that we have in human behavior of making direct connections between even avowed intention and the act itself. For most purposes, certainly, questions of intent need not be taken into account in discussions of behavior, and for some purposes it may even be useful to view the signals of animal communicators as simply manipulative, sidestepping questions as to whether they convey information or not (as, e.g. in Dawkins & Krebs, 1978). However, as long as we apply 'informational' interpretations to human communication, the 'manipulative' approach to nonhuman animal behavior involves bringing different terms to bear on the analysis of human and nonhuman animal systems; hence is unlikely to yield much understanding of continuity in the evolution of primate communication. It seems more reasonable, as a first step, to ask in what circumstances monkeys and apes appear to manipulate and mislead others, how this behavior compares with accounts of deception in other animals and whether the assumption of intention will make any significant difference to the questions we ask about behavior, or to the conclusions we arrive at concerning its evolution.

1.3. Uses of deception by nonhuman animals[2]

In any consideration of the intentions that may underlie behavior it is important first to look at the behavior. I would like to do that by asking how the deceptive behavior of monkeys and apes compares with that which has been described in other animals, and how, to begin with, we are to fit deceptive behavior into a communications theory framework. Communication can be defined in a general way as including 'any stimulus arising from one animal and eliciting a response in another' (Scott 1968). Such a definition of course ignores the systematic aspects of communication, particularly the role of feedback. Each of three basic system components – communicator, signal, and signal-recipient – is essential to communication, but, as Smith has noted, 'because communication does not occur unless there is a response, many ethologists appear to take the view that responses are the crucial components of communicating . . . [a] view [which] has tended to give ethology a preoccupation almost solely with responses and their functional implications' (Smith 1977:261). This view can be exemplified as follows:

> Communication can most conveniently be recognized empirically in terms of the relationship between the communicator's signal and the recipient's response . . . communication, *sensu stricto* necessitates the existence of a code shared between two or more individuals (Smith 1965) whose use is mutually beneficial to its possessors, i.e., increases fitness. (Klopfer & Hatch 1968:32)

I would like to distinguish between the first part of Klopfer & Hatch's statement – a rule for identifying acts of communication – and the concluding assertion, which pertains to an evolved system widespread among animals, a system common in its general character but not isomorphic in specific features (e.g. code elements) or, for that matter, functions. It is important, I think, to keep these things straight: the communicative event, the evolved underlying behavior system, and the general and specific components of each such system. Not every communicative event benefits both or all parties involved (Dawkins & Krebs 1978).[3] This is obvious in some cases of interspecific communication, an especially good example of which is seen in the sabre-toothed blenny (*Aspidontus taeniatus*) which, modelling itself in appearance and locomotion after the cleaner wrasse (*Labroides dimidiatus*), is enabled to approach larger fish closely, at which point instead of cleaning them it bites chunks out of them (Wickler 1968:162). It is less clear whether *intraspecific* lying and deception involve differential gains or losses in fitness of communicators v. signal-recipients. Here again it is important to be precise about the issues involved.

Adrian Wenner (1969), in an outline of animal behavior 'based on the origin and destination of signals', recognizes three major signal categories: intra-individual, inter-individual, and animate-environmental, of which the second (the one most relevant to a discussion of deception) is subdivided as in Table 1.1.

Interspecific signals emitted by prey species in concealment or in response to discovery by predators may contain minimal information so as not to betray the identity or whereabouts of the communicator, superabundant information that may be painful or confusing to a predator signal-recipient, or misleading information (more precisely, information that human observers tend to view as inappropriate in context). Well-known examples of arguably deceptive communication in response to discovery by predators are: distraction displays, protean displays, pursuit invitation, and mimicry of distasteful, inedible, and poisonous or otherwise dangerous species.

The several forms of mimicry have been intensively studied and debated since the English naturalist Henry W. Bates defined the initial concept over a century ago and have been reviewed in some detail by Wickler (1968). In Batesian mimicry a *mimic* species resembles a *model* species which prospective predators reject as distasteful or inedible and so conspicuously identifies itself as if to insure against their confusing it with tastier fare.

A number of inedible species may likewise resemble one another closely, and, if their mutual similarities are not the product of common ancestry and recent divergence, may have undergone convergent evolution as a measure of group economy – such at least was the

Table 1.1. *An outline of inter-individual signal categories in wider behavioral contexts*

Signal category	Behavioral context
A. *Interspecific signals*	Predation or escape from predation (including sham and mimicry) Symbiosis and parasitism Territoriality Composite schooling or flocking
B. *Intraspecific signals*	
1. Sexual	Attraction and mating Rivalry, territoriality and hierarchy
2. Nonsexual	
(*a*) Individual	Play, aggregation, and schooling or flocking Familial ties
(*b*) Colonial	Colony organization Division of labor
(*c*) Environmental	Information about food and home location Alarm and distress calls

Source: After Wenner (1969).

reasoning of the German zoologist from whom Müllerian mimicry takes its name. When similarly inedible species are similarly identified, a predator can learn to avoid all with a minimum number of bad experiences, so that the total number of individuals sacrificed in 'training' predators is spread over all species in the 'warning club' (Wickler 1968:78).

Mertensian mimicry is a concept similar in its economic assumptions and in its emphasis on learning. If potential predators are warned by a signal such as the coral pattern of both poisonous and non-poisonous snakes, the pattern adopted should be that of a species whose effect on a naive predator is intermediate, not fatally poisonous or otherwise dangerous. From the mimics' perspective, insurance is thus provided that a predator will be granted an opportunity to generalize from its bad experience of having selected a model as prey, without introducing intense selection pressure for discrimination between model and mimic. More recently, Vane-Wright (1976) and Greene & McDiarmid (1981) have argued that Mertensian mimicry is an unnecessary term, since Batesian–Müllerian concepts fully accommodate the phenomena involved.

The models of animal mimicry need not be other animal species but may be plants or inanimate objects. Insects may bear remarkable resemblances to flowers, twigs, or leaves; and color patterns which render the bearer inconspicuous against the substrate are widespread among animal species. Nor, of course, is mimicry limited to prey species. Praying mantises provide some of the most striking examples of stick-and-leaf mimicry. An 'aggressive' variant of Batesian mimicry is practiced by animals who disguise themselves as more innocuous look-alikes (e.g. the zone-tailed hawk in its resemblance to the turkey vulture: Willis 1963, cited by Smith 1977:381); who 'break the code' of communications between symbiotically related species (e.g. as mentioned, *Aspidontus*); or who lure prey by presenting themselves as prey (e.g. anglerfish: Pietsch & Grobecker 1978). 'Intended prey' is of course a figure of speech; there is no reason to suppose that intention plays a part in the formalized deception of mimicry and related animal behavior – the 'disturbing possibility' noted by Smith – and some reason not to. This is a most important point, keeping in mind that the behavior of monkeys and apes is our main concern, and I will return to it.

Wickler's review of mimicry reveals some subtle complexities in the predator–prey relationship and emphasizes the extent to which behavior of both prey and predators may be colored by deception. Partly because its explicit concern is with mimicry, but also because it deals primarily with visual signalling, Wickler's review has little to say about related important predation-avoiding strategies such as reducing signal emissions in the presence of predators or 'chorusing' of signals in confusing or even painful abundance. It does not distinguish in any systematic way between forms of deception that do and do not involve mimicry, and, in fact, its object is primarily descriptive, not classificatory. A *fine* classification of deceptive behavior would be difficult to construct; it would probably be premature as well, given the present state of our understanding of cognition.

A rough but useful categorization of deceptive behavior in the predator–prey relationship is that of Smith (1977), which deals with it from the perspective of the communicator (and without assumptions as to underlying intent): how is an animal to identify itself (as it must do for reproductive purposes and as is inevitable in conjunction with ordinary life activities) without advertising itself as prey or, if it is a predator, without frightening off prey? The behaviors which Smith says animals use to provide information without becoming prey are outlined in Table 1.2.

Note that this outline of the communicative behavior of animals that are prospective prey takes into account signal transmissions which are characterized by:

Table 1.2. *Behaviors by which animals make information available about themselves without becoming prey*

1. *Aposematic badges:* Batesian mimicry of distasteful species (conspicuousness enhanced)
2. *Crypsis:* mimicry of plants or environmental objects (conspicuousness reduced)
3. *Minimized locatability:* (focus on vocal and olfactory signals)
   (*a*) Minimal information in the physical structure of individual signals and transmissions
   (*b*) 'Confusion chorus' of signals which if emitted by a single communicator would be readily locatable
   (*c*) Chorus of signals in painful or repelling intensity, confusion apart
4. *Minimized detectability:* (focus on visual signals)
   (*a*) Minimal information as a function of time of display (e.g. 'transformation display', Morris 1956)
5. *Reactions to discovery which tend to confuse, 'cheat' or repel a predator*
   (*a*) Distraction displays
   (*b*) Invitations to premature pursuit
   (*c*) Mimicry of a frightening, more dangerous animal
   (*d*) Mobbing

Source: After Smith (1977).

1. Misleading information (categories 1 and 2)
2. Minimal information (categories 3*a* and 4*a*)
3. Superabundant information – i.e. noise (categories 3*b* and 3*c*)

Transmissions in reaction to discovery may be characterized by either misleading information or superabundant information, or may represent a mixed strategy (as arguably do distraction and mobbing displays). To be logically complete, Table 1.2 probably should include a category of responses to discovery which involve the transmission of minimal information, e.g. 'playing dead' – although that could be regarded, alternatively, as either another sort of mixed strategy or a form of mimicry which involves a same-species (but expired) model. I am not sure why Smith leaves such displays out of account. Although in my tabular summary I may have superimposed certain notions as to what constitutes systematic relations, the categorical breakdown in Table 1.2 is Smith's ; and a conceptual distinction between minimized, superabundant, and misleading information seems not only implicit in it but central to it. At any rate, consistent deployment of such distinctions in the analysis of predator–prey interactions helps clarify the relation between various forms of mimicry and between these and other behaviors which, often in conjunction with mimicry, have evolved in response to predation.

As Smith notes, 'Being inconspicuous, hard to identify, or misleading can be just as useful to a predator or a parasite as it is to its prey. There is no monopoly on formalized deception' (Smith 1977:381). The principles of deception (so to speak) for predators and parasites are consistent with those which operate in the behavior of potential prey and require no separate discussion, except to note that 'aggressive' mimicry may be more characteristic of the parasitic relation both in birds (e.g. cuckoos: Payne 1967; Wickler 1968) and in the social insects (e.g. beetles which parasitize ants: Hölldobler 1978; and slavemaking ants which parasitize other ants: Regnier & Wilson 1971; Wilson 1975) than is 'defensive' mimicry (but minimizing locatability is demonstrably effective against parasitic invasion in at least one instance: Cade 1975; 1981).

1.4. Formalized deception

Formalized behavior is Smith's term (1977:389) for behavior specialized to be informative (i.e. displays) whether it is the outcome of genetic evolution (ritualization) or changes in individual behavior (conventionalization). This paper focuses mainly on formalized behavior throughout, mostly because of the difficulty of dealing systematically in short compass with informal behavior (*vide* Birdwhistell 1970; Goffman 1974). However, there is another reason for focusing on formalized deception where the discussion is of predators and prey; it relates to the interspecific character of the relationship and of the signal category (as in Table 1.1). We are dealing here with a system in which selection pressure plays an important role, insuring that behavior specialized to be informative will be the outcome of *genetic evolution*.

The distinction between interspecific and intraspecific signals throws a new light on Klopfer & Hatch's definition of communication (see above) as involving a shared code whose use increases fitness for its possessors. It seems evident that where interspecific communication is concerned, and where the transmission of information is between predator and prey (or parasite and host), the use of a shared code is not necessarily beneficial to both or all parties involved. This raises what are clearly important issues: (*a*) whether we can speak of interspecific and intraspecific communication in the same breath and still retain a grasp of what it is we are talking about, and (*b*) whether prevarication and deception are in fact acts of communication. I hope others will agree with me that these are not issues that can be or ought to be *simply* resolved, defined 'away'.

As far as the first is concerned, perhaps what matters is whether we think of interspecific communication in terms of 'broken' rather than shared codes. With regard to the second, I am not at all sure under what circumstances *not* emitting signals might properly be called an act of communication, or whether withholding information, providing false information, and giving out superabundant and confusing information ought all to be lumped together as deceptive behavior even when intentional. What seems important from the perspective of this paper is that there exists a large and varied body of behavior which involves the manipulation of information toward the achievement of goals which human observers can perceive as goals, and that the manipulation of information raises questions of intent. For now, it is probably enough to note that there are precedents for lumping together displays which can be loosely characterized as deceptive, that the behavior involved is admittedly hard to classify, and that, as suggested, premature fine classification of deceptive behavior probably would not throw much light on cause or on underlying intent (cf. Simmons 1951).

Nice distinctions aside, the transmission of information across species boundaries is evidently not the same thing as transmission within species. It is surely significant that most descriptions of prevarication and deception in nonhuman animals involve interspecific communication. Dawkins & Krebs note that, while exaggeration and deceit ought to be profitable strategies in intraspecific competition, so far 'ethologists have failed to find many unequivocal cases of successful intraspecific deceit' (1978:302). Wickler devotes but one chapter to intraspecific mimicry, and that chapter contains but one example: mimicry by the male *Haplochromis* of eggs with his anal fin, which the mouth-brooding female attempts to collect, so that with properly timed discharge of semen the model (real) eggs in her mouth, otherwise inaccessible to spermatozoa, can be fertilized (Wickler 1968:225).

The great majority of examples of mimicry and other formalized deception pertain to predators and their prey, and an effect of natural selection in that relationship is toward ritualization of displays. It is worth noting that in the case of *Haplochromis* deception presumably contributes to fitness of both male communicator and female signal-recipient, thus meeting Klopfer & Hatch's requirements for communication, but this example of intraspecific mimicry is exceptional and, from a human standpoint, bizarre (though perhaps no more bizarre than other exceptional instances of ritualized intraspecific deception of one sex by the other: e.g. presentation of wrapped food offerings by male spiders of some species to prospective mates – presents which in some cases do not contain anything). This of course is not to say that ritualization is unimportant in intraspecific communication; obviously it is important, and it is especially important in reproductive displays. But where formalized 'deception' is concerned (again with no assumptions as to intention), the general rule appears to be ritualization of displays in interspecific communication, and in intraspecific communication a pronounced tendency toward conventionalization.

1.5. The question of devious intentions

How are we to fit nonhuman primate behavior into the framework provided by the foregoing discussion? As I will try to show, the most interesting examples of deception in monkeys and apes, while certainly not trivial from the point of view of fitness, are unrelated to stimulus-contingent fixed-action patterns of behavior on which natural selection can be assumed to bear directly; they raise theoretical problems of some interest concerning the metaphors and models of sociobiological explanations of human evolution.

Ritualization, deception, and intent

I should emphasize at the outset two main points with respect to the kinds of behavior discussed by Wickler in his analysis of mimicry and by Smith in his more general review of deceptive signalling behavior. First, most examples of animal behavior cited as deceptive involve predators and their prey, or parasites and their hosts, interspecific relationships in which the role of natural selection is direct and in many respects obvious. To the extent that we can speak of shared signals and communication between the species involved in such relationships, it is clear that displays which involve ritualized (selected for) behavior do not contribute simultaneously to the fitness of both communicator and extraspecific signal-recipients. Such displays are common among insects, less common among the vertebrates, and relatively rare among mammals (unless it is the 'featureless' aspects of some camouflaging signal structures, e.g. coat color or warning vocalizations, that is being considered). There are no good examples of interspecific mimicry by primates. The one example which Wickler gives (1968:128) has to do with the exudation by pottos of an odoriferous sebaceous substance, the main function of which presumably relates to intraspecific communication, but to which insects are said to be attracted – which the potto of course then may eat. However, and quite apart from questions concerning the context in which pottos exude an insect attractant, no model is specified by Wickler – i.e. the exudate is not compared with any particular other substance; hence it does not seem to be a good example even of mimicry.

A better example of a ritualized display employed by a nonhuman primate for deception in predator–prey interacton might be that of the male Patas monkey, who leads predators away from concealed and immobile female and immature associates – if his flight can be interpreted as a 'pursuit invitation'.

The second point is that there are far fewer examples of ritualized displays employed for purposes of *intra*specific deception, and the majority of these involve 'deception' of the female by the male in order to accomplish fertilization of eggs (as in the instances cited) or 'deception' of a signal-recipient by a communicator of the same sex through the use of inappropriate sexual signals (that is, of course, inappropriate in the eyes of human observers), as in within-sex presenting and copulatory behavior. It is clear that in instances in the first category fitness benefits are shared by both communicator and recipient. On the other hand, as far as wolf spiders and the mouth-brooding *Haplochromis* are concerned, what is involved is not an exceptional or occasional case, or even frequent cases of hoodwinking, but behavior that is the reproductive norm for the species; so, in a sense, it must be deceptive only by an interspecific stretch of the imagination.

In the second category of instances, ritualized displays associated with reproduction (presenting and mounting) are employed in contexts which appear to be nonreproductive and which do not in any way contribute directly to fertilization of eggs. Presenting by males to males and mounting of females by females are behaviors widespread among Old World monkeys; not only that, but presenting is said to be associated with 'imitation of the female genital region' by males in 'those species where the females have a colored rutting signal in the genital zone' (Wickler 1972: 209–10). Here the prototypical example is the Hamadryas baboon, males of which species resemble females more in color and degree of perianal swelling than do, for instance, males of *Papio anubis* resemble females of that species. It is not at all clear from a social functional standpoint why there should be such differences between these species, but according to Wickler it constitutes

> intra-specific morphological signal-imitation . . . [in which] the primary signal is derived from the sexual behavior and has been made available for social functions by alterations in function and motivation . . . . It may be that the signal-structures arose in females as a 'by-product' of physiological processes but the development of analogous (and not homologous) structures in males would seem to be due to their specialized evolution as a signal apparatus . . .[4] (Wickler 1967:98)

> The general concept is that the male's conspicuous hindquarters have, by a process of adaptation, become closely similar to the sexual skin of the female in order to elicit in a conspecific a reaction originally directed towards a female. The signal-receiver is, therefore, in a certain sense deceived. (Wickler 1967:122).

Wickler and others have made much of this postulated transformation of signal functions from a sexual reproductive context to a within-sex aggressive-dominance system function. The theory played an important part in the perpetuation and popularization of an old primate behavior model that overemphasized and oversimplified links between male-male competition for females, aggressive dominance, defense against predators, and territorial behavior. From the standpoint of the present discussion the main question is not the extent to which there has or has not been a functional transformation of a ritualized display, but whether or not in nonreproductive presenting and mounting there is in fact transmission of information which confuses or misleads signal-recipients in any meaningful sense. This is a difficult question to answer as long as the specific goals of signal-recipients are unknown, which they ordinarily are in natural circumstances.

In any event, none of this behavior, and none of these displays which in their primary functions bear an obvious and direct relation to fitness (and so might be grist for the mill of behavior selection) seem to fit definitions of deception that hinge on intent – and this is apart from the issue of *demonstrating* intent. In all the foregoing examples (except the last, where the question is not so much whether there may be underlying intention as whether deception is involved, and what the goals of signal-recipients may be) what is important is that no assumptions as to intent are required. Indeed, in the ritualization of 'deceptive' behavior, the workings of selection would seem to be impeded rather than furthered by anything approaching self-conscious reflection or ruminating on the model by some artist of mimicry. If monkeys and apes have devious intentions they are presumably not focused on the strategic deployment of ritualized displays (though see Hrdy 1979). This is an important point, I think, because if it is correct it breaks a necessary link between metaphor and natural reality in one of the major models of sociobiology.

1.6. Conventionalized deceptive behavior in monkeys and apes

Most reports of deception by monkeys or apes, whether of conspecific companions or of organisms of other species (e.g. human observers), involve signals which appear to be the product of conventionalization rather than ritualization.[5] Such reports tend to be anecdotal, especially where behavior in natural circumstances is concerned, for what are perhaps obvious reasons: in natural circumstances, deception which was tied consistently to goals recognizable by human observers, and which therefore lent itself to systematic investigation, would be likely to train any but the densest signal-recipients to ignore it before those investigations could get under way!

One category of conventionalized deception from which I have already drawn examples is that of prevarication in gorilla and chimpanzee language-learning studies. These commonly involve what are purported to be joking misrepresentations of fact or attempts to conceal or blame others for responsibility of some action committed by the communicator. I say 'purported to be' because displays in these instances involve signs imposed or acquired from outside the known 'natural' range of species-specific behavior, giving rise to significant problems of meaning and interpretation. Since both the examples and the problems are well-known I will not discuss them further, except to note that (*a*) since language-learning studies ordinarily have involved training animals to use signals in straightforward transmission of information (that is, not in experimentally structured attempts to conceal information or to mislead) reports concerning the use of those signals or variants thereof in *deception* involve unanticipated departures from systematic procedure and are anecdotal; and (*b*) interpretation of the meaning of such departures has usually been confounded by the fact that they are reported by trainers who assigned original meaning to signals imposed or acquired and whose judgements as to whether subsequent use of such signals represents misunderstanding or intentional misuse cannot be assumed to be unbiased.

This is not to say that other anecdotal accounts of deception in nonhuman primate behavior are necessarily more informative or more reliable, only that they may not be freighted with quite the same weight of bias as are reports by those who perceive intentional deception in the use of signals to which they have made prior definitional contributions and of which they subsequently are signal-recipients (and who, to further confound the issue of 'misuse' v. 'misunderstanding' have themselves trained ape communicators in the 'correct' use of said signals)!

In natural circumstances, apart from isolated incidents (van Lawick-Goodall's account (1971:107) of young Figan ignoring a banana until Goliath had left the area), or reports on limited series of observations (Hrdy's report (1979:33) of pseudo-estrus in Langurs), there are at least two well-reported areas of behavior in which premeditation and subterfuge appear to play a role. One of these is infant stealing, especially when a preliminary bout of grooming serves to relax the mother and render the infant more accessible (Hrdy 1976:138). The other is hunting as practiced by chimpanzees and as described by van Lawick-Goodall (1971) and, in some detail, by Teleki (1973). Van Lawick-Goodall and especially Telecki have explicitly emphasized the importance of cooperative strategic maneuvering, decision-making, and premeditation in chimpanzee hunting (deception *per se* is not always a central element), and these terms seem to be justified, though one might easily argue that it all could be described equally well in terms of heightened concentration, receptive integration of a variety of closely clustered or simultaneous signals, and coordination of individual activities; in short, in terms consistent with modern interactional theory such as that of McBride (1975).[6]

One returns always to the question of how to establish the premeditated intent, devious or no, that may underlie action. The reason that deceptive actions appear to be relevant to this problem is that they evidently involve a conflict between what appears to be the goal of action, and what turns out to have been the communicator's 'real' goal. Most reports of deception by monkeys and apes in natural circumstances necessarily involve isolated incidents, and one cannot but distrust anecdotal accounts in which the assessment of intention is dependent upon an observer's ability retrospectively to distinguish clearly between at least three possibilities: (*a*) intentional deception by the communicator; (*b*) misunderstanding of goals by the signal-recipient (no deception or devious intent assumed); and (*c*) correct statement but subsequent substitution of goals by the communicator (again no deception or devious intent assumed).

The problem of how to correctly identify and reject possibilities alternative to intentional deception seems obvious. One could take *any* anecdotal account of devious intention and deceptive action in monkey or ape behavior and interpret it in a way that made the postulation of intent superfluous. For instance, to take a problematical example from my own experience, I once fell out of a tree when a rhesus monkey passing beneath me suddenly bounded into it and over my head for a 'branch-shaking' display. I had anticipated some such action because as he came down the path he had too obviously ignored all my throat-clearings and vague shiftings of limbs – polite attempts to alert him to my presence (I was having a lunch break, not trying to blend in with the scenery). Distrusting this refusal to acknowledge my signals, I was easing myself into a less precarious position when he sprang into the tree, hence my fall. Most field primatologists no doubt can recall like episodes in which individuals seemed to ignore their presence in order to mask some plan of action in which, as events proved, they were to figure prominently.

Alternatively, of course, this incident could be interpreted as follows: a rhesus monkey, alerted by some vocal noises to the presence of an obtrusive human being, bounded into a tree and gave a branch-shaking display. Since that is pretty much what did happen, since I prefer in the ordinary way to deal with the meaning of behavior in terms of signals and responses that can be clearly and operationally defined, and since for most purposes understanding of behavioral events is not advanced by assumptions concerning *a priori* awareness and planned action – for these reasons my habit has been to tuck such incidents away in memory's store and not put them forward as evidence of underlying intention. Nor am I putting this one forward now as such evidence, only as an example of how difficult we may find it to substantiate assertions as to intention.

Even when the evidence is more than anecdotal, when, as in infant-stealing and hunting, repeated observations have established recurrent patterns which provide a basis for something more than intuitive analysis of central components, the problem of alternative interpretations is hardly lessened. Suppose the following sequence of interactions is observed in, e.g. a troop of rhesus monkeys: a young female approaches a mother and infant, initiates grooming of the mother, by degrees transfers her grooming activities to the infant, then suddenly scoops up the infant and runs off. A human observer of such a sequence might very well decide that there were features in the groomer's behavior which, in retrospect, seemed to have foreshadowed the event: a hesitation in the approach, perhaps, a divided attention during the interaction, a perfunctoriness in the grooming activity. The observer might then decide to look for these same features in other series of grooming bouts that involved (for ideal purposes of control) the same three individuals.

For alloparent (A), biological parent (B), and infant (C), our observer might accumulate data on three kinds of encounters:

1. A approaches and grooms B but not C and departs
2. A approaches and grooms B and C and departs
3. A approaches and grooms B and C and kidnaps C

If comparison of such hypothetical grooming bouts revealed clear-cut differences in A's behavior in the third series as opposed to the first and, especially, the second, would those differences argue premeditated kidnapping? They might, but the point of this 'for-instancing', and the reason that the instancing has to be hypothetical, is that the kind of clear-cut behavior differences that would lend themselves to such neat comparisons would be most unlikely to emerge under natural circumstances. For, unless one assumes that monkeys (and apes) are somehow less sensitive interpreters than are human observers of the signals emitted by conspecifics (including unconscious signals that might betray underlying devious intent), then parent B (whom we should allow the same capacity for reflection as alloparent A) would presumably respond to any but the most subtle departures from A's customary grooming style by anticipating theft of her infant and increasing her protectiveness. The net effect of learning and wariness would be to increase similarity of signaling behavior at least between series 2 and series 3.

If any lesson can be drawn from such speculations it is that conventionalized displays which lent themselves to premeditated deception would have to be subtly variable if the same individuals were to be victimized more than infrequently, in order to avoid training them not to be deceived. Thus, I believe, the reason we mostly get anecdotal accounts of deception, at least in natural circumstances, with no convincing recurrent series of observations, is a sampling problem which reflects a more interesting methodological problem: how to distinguish between misunderstanding and misuse of signals when factors of learning are almost certain to contaminate sample series. I *think* this is a real problem. One way to get around it would be to insure that victims did not learn from being gulled (and so did not train perpetrators of deceit into more and more subtle forms). This can be done experimentally by using human stooges as 'victims' in cross-specific deception and enjoining them against learning. This has in fact been done (Woodruff & Premack 1979) in a series of experiments testing the ability of chimpanzees to convey and to comprehend accurate and misleading information concerning the location of hidden food. In one test a chimpanzee was informed of the location of food but denied access to it:

> The animal could obtain food only by imparting information about its location to an uninformed human positioned outside the enclosure, in the vicinity of the goal. One human was friendly and cooperative; if he found the food he gave it to the chimpanzee, but if he failed the animal received nothing. Another human was hostile and competitive; if he found the food he kept it for himself, but if he failed the chimpanzee was allowed to leave the enclosure and obtain the food. (Woodruff & Premack 1979:335)

In a second test roles were reversed. Humans informed about the location of food modeled signals after those of chimpanzee communicators in the first test to indicate either the correct location or, in the case of competitive trainers, an incorrect location.

In the first test situation four chimpanzee subjects quickly learned to indicate to the cooperative trainer the correct location of food, progressing toward more organized signaling behavior and toward more similarities in their signals: all four eventually used an outstretched arm or leg to point to the baited container. Working with the competitive trainer, however, they suppressed information; and two of the four learned to convey misleading information, pointing to the unbaited container with significant regularity and in contradistinction to their behavior with the cooperative trainer. It is here that, for purposes of training, the human accomplices were enjoined from learning:

> Each trainer was urged to develop an accurate choice strategy and then maintain that strategy as much as possible; if a subject later began to mislead him on repeated trials, he did not then change his strategy to outwit the subject (e.g., by choosing the opposite container). (Woodruff & Premack 1979:338).

In the second test situation, with no such restrictions imposed on chimpanzee subjects, three of the four chimpanzees ultimately learned to controvert the competitive trainer's signals by avoiding the container toward which he oriented (ibid p. 354).

In Woodruff & Premack's experiments, the two subjects who consistently misinformed the competitive trainer used relatively gross, invariant signals for deceptive purposes (the same conventionalized pointing displays with which they imparted to cooperative trainers correct information); but in view of the methodological problem raised above, I would like to reemphasize that the human stooges 'victimized' by such behavior were expressly required to be consistent in their responses, i.e. not to 'see through' blatant lies and reform behavior accordingly. It may be that the relation between devious intentions, deceptive action, and response to deception are not accessible to the kind of descriptive and analytic procedures we are accustomed to applying to natural social behavior. At any rate it seems that, in experiments which bear on deception and underlying intention, either the experimental structure must be loose enough to accommodate relatively free interplay among conspecifics (as in Menzel's study), in which case the problem of distinguishing between misuse v. misunderstanding of signals (or between concealment of information v. forgetting) cannot easily be resolved; or else, to get around that problem, strict controls have to be established by human participants who (to maintain standard signal emission) end up cheating a bit on the issue of training. That suggestion of course is not intended to detract from the importance of experimental work. Woodruff & Premack maintain that, in the case of at least two of the subjects (who consistently misinformed the competitive trainer), 'these instances of deceit meet the most stringent behavioral criteria for intentional communication'. Not everyone will agree with their claim, certainly, but it may well be that in their work and, more especially perhaps, in Menzel's we come as close as possible to experimental demonstration of devious intentions in apes.

It surely is significant that almost all accounts of deception in monkey and ape behavior deal with concealment of information, and that such concealment is viewed as deceptive by human observers who share with nonhuman primate communicators information of which signal-recipients are in the dark, with little or no opportunity to learn that they are being deceived. The important exception is the experiment of Woodruff & Premack in which human 'victims' of deception know that they may be 'victims' but are prohibited from taking advantage of that knowledge to learn appropriate responses. This suggests, to me at least, that just as natural selection does not appear to affect, via ritualization, that behavior of nonhuman primates which may be (but cannot be 'proven' to be) linked to devious intentions, so ordinary learning processes would appear to constitute sufficient defense against deceit – so that no special genetically evolved defenses need be called into play. Sociobiological explanations of human culture which postulate genetic evolution of ethics in a context of reciprocal altruism are interesting essays in logic which can be very useful in teasing out some relations between genetic and behavior systems, but in no way is their logic a direct expression of natural relations.

So far I have largely omitted from discussion that category of deceptive behavior most frequently cited as evidence of intention, premeditation, and even creativity (Hayes 1978:216). Description of the 'lunch-games' of laboratory chimpanzees, practical jokes played on spectators by zoo animals, and other boredom-dispelling pastimes of monkeys and apes held captive in limited environments tend to be couched in anthropomorphic and excessively cute terms, no doubt because of the contexts in which such behavior is observed. They are not the less interesting for that, and the activities which they describe are significant in spite of – I would say because of – their evidently trivial character.

An example of such activities is 'chicken-teasing', a game invented by the chimpanzees of Köhler's Tenerife colony (Köhler 1925). The game had two variant forms, and I quote Hayes's concise description:

> In its first form the ape holds out a piece of bread to the chickens who feed just outside the apes' large exercise cage. When a chicken approaches and trustingly pecks at the bread, the ape whisks it away, leaving the chicken with a healthy bite of air. Since the chickens never achieve a full appreciation of their role in the game, it can be repeated as many as 50 times in the course of a jolly ape lunchtime.
>
> In the rough form of the game, the ape holds the bread in one hand and a stick in the other. When the chicken approaches to eat, he gets a sharp poke in the feathers. (Hayes 1978:216–17)

The second of these games, albeit harder on the chicken, is probably more interesting from the apes' standpoint, and certainly from mine here, for it necessitates more in the way of strategic manipulation of behavior components to keep the chicken in play. Chimpanzees in open cages are notorious for playing similar tricks on humans, using whatever materials lie at hand (feces being one such). Chimpanzees in the Yerkes Orange Park laboratory became so adept at dousing passersby with drinking water, that even visitors warned in advance seem to have been easy marks. Hediger's relation of how he was drenched despite his own considerable experience with apes, explicit forewarning, and precautions taken, makes a delightful cautionary tale:

> He sat there listlessly, with his back to the cage bars, and seemed to be playing with his toes. . . I looked at his cheeks, which didn't seem to me to be the least bit swollen. So, as if by chance, I passed between his cage and the bamboo grove, keeping as far as possible – some twelve feet – from the cage.
>
> When I had reached the narrowest part, the old chimpanzee suddenly swung round in a flash, reaching the railings in one leap, and, unable as I was to retreat any further, drenched me from head to foot in a stream of warm water. It gushed out as from a hydrant. The cunning ape must have been watching me for some time out of the corners of his eyes, and have taken in his water supply as a precaution; he then squatted hypocritically, acting the innocent so well that he caught me out a hundred percent at the spot most favorable to him. (Hediger 1955:134)

Although human beings may be warier than chickens, traps sprung on them may not require baiting. Monkeys and apes are bait enough in themselves; or, if not, human beings frequently can be lured close through the initiation of eye contact or some flattering postural–gestural sign of recognition or 'interest'. The Basel Zoo's gorilla, Achille, acquired the habit when he desired human company of pushing his arm through the top of the wire mesh of his cage and 'pretending' to struggle to extricate it (Hediger 1955:150).

For the past several months my students and I have been working with gorillas in the Denver Zoo, monitoring social behavioral indicators of physiological events in the female reproductive cycle. All three of the Denver Zoo's gorillas (two females, Bibi and Maguba, and one male, Thomas) occasionally direct toward spectators displays which contain variable and, for all three, roughly similar components of chest-pounding, vocalization, locomotion, and kicking or slamming one of the Plexiglas windows that wall off spectators from the indoor portion of the gorilla habitat. Maguba's displays, unlike those of either Bibi or Thomas, often occur not only in extended but in greatly varied sequences which are highly effective in capturing and maintaining spectators' attention, and which she employs to manipulate responsive individuals up and down the viewing area. The central component is a single slam of the window with one hand, which almost invariably elicits a startled reaction in spectators. It would not take many repetitions of this, obviously, until a spectator learned to predict it and ceased to experience that delicious thrill of surprise and shock the elicitation of which is apparently Maguba's goal. But through flexible variation of accompanying signals, adroit timing, and above all deception, Maguba makes it difficult to predict her climactic window slam (anti-climactic, actually, for it is almost always followed by a run to a new location). Deception seems to be important in two ways here: (*a*) a seeming pretence of indifference by Maguba to signal-recipients' response readiness, as she looks in every other direction, monitoring spectators' vigilance peripherally; and (*b*) when, after a few hit-and-run performances, spectators manifest reluctance to be lulled again into relaxation, keeping their attention firmly fixed on her, she will 'fake' a window slam, stopping her arm at the last moment.

This is, I think, not at all uncommon behavior for captive nonhuman primates. When we speak of caged animals dispelling boredom or breaking monotony, what we presumably mean is that they act to vary or increase sensory input. This can be done in any number of ways: eating, allogrooming, 'fiddling with' inanimate objects, etc. Or it can be accomplished through social interaction, either simply repetitive (getting a chicken to peck at air over and over again), or complexly varied. Kohler's chimpanzees in the second variant of chicken-teasing, Yerkes' chimpanzees with their water games, and the Denver Zoo's Maguba all seem to be doing the same thing – provoking startle reactions for the sake of observing them. In each case, the limiting circumstances of the game (confinement), require deception so that the game may be continued and enjoyed.

Such stereotypical tricks may be played as well on conspecific companions in confined circumstances. In a group of enclosed rhesus monkeys that I observed on Parguera Island, Puerto Rico, some years ago, a young male persistently bedevilled subordinate companions by pinching them or pulling sharply at their pelage in the middle of grooming bouts, again for no apparent purpose but to observe their reactions. Naturally enough, his victims were reluctant to submit to such indignities; nevertheless, I was surprised at the frequency with which he could, by persistent overtures, continue to initiate grooming that all too often ended with a pinch or a yank. It is difficult to imagine that behavior so apparently objectionable could maintain itself in such repetitious form in natural circumstances, though of course surprising one's companions through more variable if not more subtle *ad hoc* deceptions are an important element in monkey, ape, and human play. It is possible that in confinement, with possibilities for action limited to begin with, animals are more likely to lend themselves to deception, i.e. are less likely to distinguish between 'straightforward' and 'devious' intentions, even when failing to do so subjects them to periodic minor abuse. Such speculation seems not entirely unwarranted. Steve Maier's elegant studies of learned helplessness (Maier & Seligman 1976; Maier & Jackson 1979), a learning deficit phenomenon widely reported in experimental subjects and animals in confinement, suggest that animals subjected to uncontrollable events develop something like a general proposition regarding the effectiveness or ineffectiveness of behavior.

1.7. Review and discussion

A communicative event involves a communicator or sender of signals, the signals themselves, and a signal-recipient. Content is implied; i.e. signals convey information, and the information may be misunderstood or misused (but cf. Dawkins & Krebs 1978:286–7). *Deception* is a term with a very broad referential base, but we usually apply it to communicative events in which information is intentionally misused: we are misled by incompetent leaders, but we are not deceived by them unless they lie to us in the process; which of course they may do by not advertising incompetence of which they are aware (unless failure to broadcast information about one's shortcomings is an inherent feature of incompetent leadership – it should be clear by now that deception is hardly less slippery a concept than intention). It is because intention is assumed to underlie deception that deception is so often presented, explicitly or implicitly, as evidence of intention.

According to Woodruff and Premack:

> Communication is intentional if . . . the sender (i) appreciates the fact that his behavior transmits information, (ii) recognizes that the recipient also knows that his behavior is informative, and (iii) is able to choose from a set of alternatives that course of action (or inaction) which will provide (or suppress) a given bit of information. Intentional communication is thus more than a simple transfer of information; it is a purposive transfer, based on the sender's knowledge about the effect that his actions can have on the recipient. (Woodruff & Premack 1979:334)

By this definition much of what is sometimes loosely subsumed under deception in animal behavior is not deception at all. In mimicry and most related behaviors it is clear that intention is not a factor, and in none of these genetically evolved displays does the postulation of intention contribute to understanding of the behavior in context. Most such displays involve interspecific communication and selection pressure for relatively invariant signal characteristics.

Smith makes a useful distinction between ritualized displays (products of genetic events) and conventionalized displays (in which learning plays a dominant role).[7] References to deception in monkey and ape behavior have usually involved conventionalized displays, and behavior which is purported to reflect underlying devious intentions always involves conventionalized displays. This may be a matter of definition, though it is not clear that it must be.

That social strategies of monkeys and apes may be premeditated is suggested by both naturalistic observations and experimental research. Experiments by Menzel and by Woodruff & Premack are especially convincing, and these may come as close as can be done to confirming the operation of devious intentions in nonhuman animals. Obtaining a series of observations of deceptive behavior depends on recognition by human observers of signal or context features that betray devious intentions and misuse of information. This raises problems of sample contamination by learning for, if such signal/context features are consistently recognizable, intended signal-recipients should learn to recognize them and take counter-measures against deception. In Menzel's experiments conditions perhaps were structured loosely enough that this did not occur; however, and as a consequence, there remains a problem of distinguishing between misuse and misunderstanding of information. In Woodruff & Premack's experiments, that problem was avoided by not allowing human stooges to learn from their experience when deceived by chimpanzee subjects repeatedly and under relatively invariant conditions. Stereotypical patterns of actively misleading behavior, such as those employed by two of Woodruff & Premack's chimpanzees, are uncommon in the wild (most anecdotal accounts of deception in natural circumstances involve simple withholding of information), but the 'tricks' which confined monkeys and apes play on humans and on each other, which elicit startle reactions and increase sensory input, may be interpreted as analogous results of something like 'natural experiments'.

Smith notes the 'disturbing possibility' that nonhuman animals may manipulate displays to deceive and says that there is so far no evidence that they do that. A review of deceptive behavior in monkeys and apes convinces me that those animals do manipulate displays in complex and subtle ways; however, (*a*) in intraspecific interactions in natural circumstances there is seldom any need for nonhuman primates to mask immediate intentions (and there is probably no way that we can demonstrate relations which we take to exist between immediate goals and long-range strategies) so that the question of devious intentions is largely irrelevant to the methods and problems of naturalistic field studies; and (*b*) displays that are manipulated are in any event conventionalized, so that they can be more meaningfully discussed in terms of learning or, as in human behavior, cultural constraints, not selection.

This, I believe, has important implications for evolution theory. It is unlikely that animals, including monkeys and apes, can 'detach' themselves sufficiently from displays which have been standardized by natural selection and adaptation to manipulate them like chess pieces. Therefore, references to *ad hoc* genetic cost/benefit accounting in ongoing behavior must always be metaphorical. Furthermore, and apart from that all too familiar caveat, it is difficult to see how the evolution of human ethical or other cultural behavior systems can be very usefully accounted for in terms of a see-saw between deception and discrimination (and refined deception and further refined discrimination) in which selection is assumed to play any sort of direct role. There seems to be a basic confusion in explanations such as the following of how reciprocal altruism might produce a system of ethics:

> During the course of natural selection, individuals are certain to arise who (unconsciously) try to 'cheat' on the altruism game. They may produce sham moralistic aggression, sham guilt, sham sympathy, and sham gratitude, in an attempt to take more than they give – a situation that could be biologically beneficial in the short

> term, at least. Although an animal without spoken language could indulge in such cheating, it is much more effectively disguised by means of the spoken word.
>
> Just as natural selection inevitably produces would-be cheaters, it will inevitably give rise to individuals capable of detecting cheating. And so the game of bluff and double-bluff begins, with the new emotions of trust and suspicion being invented. (Leakey & Lewin 1978)

The confusion in such accounts lies in their suggesting that behavior is 'unconscious' with respect to goals, while proceeding to discuss it wholly in terms of premeditation and purpose. As far as conventionalized deception is concerned, reductionist response-focused evolutionary parables are simply irrelevant to the complex interaction of individual developmental and communications feedback processes, in their influence on both the structure and the interpretation of displays. Since ritualization appears to play so unimportant a role in the production of deceptive displays by monkeys and apes it seems unlikely that those animals, at least, could ever evolve ethical systems through anything like the direct action of natural selection.

There probably is room for both natural selection and devious intention in evolution theory, but perhaps not at the same level of analysis. To try to extract information about inter-generational change through the application of models of individual learning/decision-making to problems of behavioral ecology is in some ways a strange exercise, one that seems to me unlikely to prove successful. Our best hope of understanding the biological evolutionary bases of ethical systems, and of human culture in general, probably lies in conventional analytic theory building. Working from observed instances of behavior through inferred constructs to different observed instances, continually improving our understanding of relations between (for instance) games logic, cognitive processes, and interactional possibilities in specific settings, we may one day gain analytic access to the ultimate causes of complex social behavior. At any rate, in my opinion, only this analysis and synthesis of formal theory will allow us to elaborate the details of how natural selection shapes behavior in evolution.

Notes

1. I.e., in either direction. In some ways chimpanzee-learning studies are most interesting as poignant failures by *Homo sapiens* to communicate to its closest relatives how we process information in our econiche.
2. As commonly used, the term *deception* sometimes implies underlying

intention, sometimes not. I use it in both senses, usually relying on context to make clear the particular sense in which it is used.

3. Dawkins & Krebs define communication as something that occurs 'when an animal, the actor, does something which appears to be the result of selection to influence the sense organs of another animal, the reactor, so that the reactor's behaviour changes to the advantage of the actor' (p. 283). In their view, any benefit to the reactor is coincidental, and they explicitly reject 'informational' interpretations of communication. Their 'manipulative' interpretation may be itself no less biased than the one against which it is set. Dawkins & Krebs emphasize the energetic efficiency of signal transmission from one point of view, that of the actor, who they say is thereby enabled to put to work to his own purposes the physiology and muscle power of reactors. But signal reception can work in a similar way to the advantage of the reactor, enabling him to utilize the actor as a second, spatially detached, sensory data-processing system. It is really only by refusing to take into account the perspective of the signal recipient that Dawkins & Krebs are enabled to question the utility of the concept of information to models of animal communication. They note that 'most informational interpretations of animal communication have concentrated on information about the actor's internal state rather than about events in the real world' (p. 287). Such a distinction, it seems to me, must be irrelevant from the point of view of the signal recipient, much of whose behavior can only finally be comprehended in terms of sensory input of many sorts, including pre-processed data in signals from other organisms. Whether we are to regard this as information about the external environment or only about the internal environment of an external data-processing system seems entirely a question of point of view.

   Nevertheless, despite these objections, and apart too from its uses as antidote against anthropomorphic models of signalling behavior, Dawkins & Krebs' paper contains a wealth of insights into the nature of communication, and I wish I had encountered it before writing this one, instead of belatedly afterward as was the case. (I thank Emil Menzel for directing me to it.) We differ considerably, I think, in our view of the evolution of behavior as well as in that of the nature of communication; but there are close parallels in our discussions of deception, especially concerning the limiting effects of learning on the continuation and spread of intraspecific deceit (cf. pp. 302–4, Dawkins & Krebs).
4. The distinction between 'analogous' and 'homologous' structures seems to me generally less helpful than intended, particularly so when, as in this case, the reference is to structures of conspecific organisms.
5. I use the term *conventionalization*, following Smith (1977:10), for 'processes of specialization based on individual experience,' i.e. learning, as opposed to processes involving genetic evolution, i.e. ritualization.
6. In which the unit of analysis of behavior 'is not two separate animals making behavior towards each other but the whole single interaction, in which the behavior of both must be treated together, as each responds to the other at all times' (McBride 1975:329).
7. Cf. Kummer's 'phylogenetic adaptation' and 'adaptive modification' (Kummer 1971).

References

Birdwhistell, R. L. (1970). *Kinesics and Context: Essays on Body Motion and Communication.* Philadelphia: University of Pennsylvania Press.

Cade, W. (1975) Acoustically orienting parasitoids: fly phonotaxis to cricket song. *Science*, **190**, 1312–13.
Cade, W. (1981). Alternative male strategies: genetic differences in crickets. *Science*, **212**, 563–4.
Chalmers, N. (1980). *Social Behaviour in Primates*. Baltimore: University Park Press.
Dawkins, R. & Krebs, J. R. (1978). Animal signals: information or manipulation? In *Behavioural Ecology, an Evolutionary Approach*, ed. J. R. Krebs & N. B. Davies. Oxford: Blackwell Scientific Publications.
Ferguson, E. S. (1977). The mind's eye: nonverbal thought in technology. *Science*, **197**, 827–36.
Gardner, R. A. & Gardner, B. T. (1971). Two-way communication with an infant chimpanzee. In *Behavior of Non-human Primates*, ed. A. M. Schrier & F. Stollnitz. New York: Academic press.
Geschwind, N. (1971). Some differences between human and other primate brains. In *Cognitive Processes of Nonhuman Primates*, ed. L. E. Jarrard. New York: Academic Press.
Goffman, E. (1974) *Frame Analysis: an Essay on the Organization of Experience*. New York: Harper & Row.
Greene, H. & McDiarmid, R. W. (1981). Coral snake mimicry: does it occur? *Science*, **213**, 1207–12.
Gregg, L. W. (1971). Similarities in the cognitive processes of monkeys and man. In *Cognitive Processes of Nonhuman Primates*, ed. L. E. Jarrard. New York: Academic Press.
Griffin, D. R. (1976). *The Question of Animal Awareness, Evolutionary Continuity of Mental Experience*. New York: Rockefeller University Press.
Hayes, J. R. (1978). *Cognitive Psychology: Thinking and Creating*. Homewood, III.: The Dorsey Press.
Hediger, H. (1955) *Studies of the Psychology and Behaviour of Captive Animals in Zoos and Circuses*. Translated by G. Sircom. New York: Criteria Books.
Hewes, G. W. (1973). Primate communication and the gestural origin of language. *Current Anthropology*, **14**, 5–24.
Hewes, G. W. (1977). A model for language evolution. *Sign Language Studies*, **15**, 97–168.
Hölldobler, B. (1978). Ethological aspects of chemical communication in ants. *Advances in the Study of Behavior*, **8**, 75–115.
Honig, W. K. (1978). On the conceptual nature of cognitive terms: an initial essay. In *Cognitive Processes in Animal Behavior*, ed. S. W. Hulse, H. Fowler & W. K. Honig. New York: Wiley.
Hrdy, S. B. (1976). Care and exploitation of nonhuman primate infants by conspecifics other than the mother. *Advances in the Study of Behavior*, **6**, 101–58.
Hrdy, S. B. (1979). Infanticide among animals. *Ethology and Sociobiology*, **1**, 13–40.
Hulse, S. H., Fowler, H. & Honig, W. K. (1978). *Cognitive Process in Animal Behavior*. New York: Wiley.
Klopfer, P. H. & Hatch, J. J. (1968). Experimental considerations. In *Animal Communication, Techniques of Study and Results of Research*, ed. T. A. Sebeok. Bloomington: Indiana University Press.
Knapp, M. L. & Comadena, M. E. (1979). Telling it like it isn't: a review of theory and research on deceptive communications. *Human Communication Research*, **5**, 270–85.
Köhler, W. (1925). *The Mentality of Apes*. New York: Harcourt.
Kummer, H. (1971). *Primate Societies: Group Techniques of Ecological Adaptation*. New York: Aldine-Atherton.

Lawick-Goodall, J. van. (1971). *In the Shadow of Man*. Boston: Houghton-Mifflin.
Leakey, R. & Lewin, R. (1978). *People of the Lake*. New York: Doubleday.
Leonard, J. W. (1979). A strategy approach to the study of primate dominance behaviour. *Behavioural Processes*, **4**, 155–72.
Livingstone, F. B. (1973). Did the Australopithecines sing? *Current Anthropology*, **14**, 25–9.
McBride, G. (1975). Interactions and the control of behavior. In *Socialization and Communication in Primary Groups*. The Hague: Moulton.
McCall, G. J. & Simmons, J. L. (1978). *Identities and Interactions*, revised edn. New York: The Free Press.
McGuigan, F. J. (1978). *Cognitive Psychophysiology: Principles of Covert Behavior*. Englewood Cliffs, NJ: Prentice-Hall.
McGuigan, F. J. (1979). *Psychophysiological Measurement of Covert Behavior*. New York: Wiley.
Maier, S. F. & Jackson, R. L. (1979). Learned helplessness: all of us were right (and wrong): inescapable shock has multiple effects. *The Psychology of Learning and Motivation*, **13**, 155–218.
Maier, S. F. & Seligman, M. E. P. (1976) Learned helplessness: theory and evidence. *Journal of Experimental Psychology*, **105;** 3–46.
Menzel, E. W. (1978). Cognitive mapping in chimpanzees. In *Cognitive Processes in Animal Behavior*, ed. S. H. Hulse *et al.* New York: Wiley.
Morris, D. (1956). The feather postures of birds, and the problem of the origin of social signals. *Behaviour*, **9**, 75–113.
Patterson, F. (1980). In search of man: experiments in primate communication. *Michigan Quarterly Review*, **19** (1), 95–114.
Payne, R. B. (1967). Interspecific communication signals in parasitic birds. *American Naturalist*, **101**, 363–75.
Pietsch, T. W. & Grobecker, D. B. (1978). The compleat angler: aggressive mimicry in an Antennariid Anglerfish. *Science*, **201**, 369–70.
Premack, A. J. & Premack, D. (1972). Teaching language to an ape. *Scientific American*, **227**, 92–9.
Quiatt, D. (1966). *Social Dynamics of Rhesus Monkey Groups*. Ann Arbor: University Microfilms.
Quiatt, D. (1979). Aunts and mothers: adaptive implications of allo-maternal behavior of nonhuman primates. *American Anthropologist*, **81**, 310–19.
Regnier, F. E. & Wilson E. O. (1971). Chemical communication and 'propaganda' in slave-maker ants. *Science*, **172**, 267–9.
Rumbaugh, D. M., Glasserfeld, W. von & Pisani, G. (1973). Lana (chimpanzee) learning language: a progress report. *Brain and Language*, **1**, 205–12.
Scott, J. P. (1968). Observation. In *Animal Communication, Techniques of Study and Results of Research*, ed. T. A. Sebeok. Bloomington: Indiana University Press.
Simmons, K. E. L. (1951). The nature of the predator-reactions of breeding birds. *Behaviour*, **4**, 161–71.
Smith, W. J. (1965). Message, meaning and context in ethology. *American Naturalist*, **99**, 405–10.
Smith, W. J. (1977). *The Behavior of Communication, an Ethological Approach*. Cambridge, Mass.: Harvard University Press.
Teleki, G. (1973). *The Predatory Behavior of Wild Chimpanzees*. Lewisburg: Bucknell University Press.
Vane-Wright, R. I. (1976). A unified classification of mimetic resemblances *Biological Journal of the Linnean Society*, **8**, 25–56.
Wenner, A. M. (1969). The study of animal communication: an overview. In

Approaches to Animal Communication, ed. T. A. Sebeok & A. Ramsay, The Hague: Mouton.
Wickler, W. (1967). Socio-sexual signals and their intra-specific imitation among primates. In *Primate Ethology*, ed. D. Morris, Chicago: Aldine.
Wickler, W. (1968). *Mimicry in Plants and Animals*. New York: McGraw-Hill.
Wickler, W. (1972). *The Sexual Code; the Social Behaviour of Animals and Men*. Garden City, NY: Doubleday & Co.
Willis, E. O. (1963). Is the zone-tailed hawk a mimic of the turkey vulture? *Condor*, **65**, 313–17.
Wilson, E. O. (1975). *Sociobiology, the New Synthesis*. Cambridge, Mass.: Harvard University Press.
Woodruff, G. & Premack, D. (1979). Intentional communication in the chimpanzee: the development of deception. *Cognition*, **7**, 333–62.

COMMENT

ROBERT M. SEYFARTH

Quiatt's excellent review raises a fundamental question in the analysis of animal communication, particularly that of the supposedly 'deceitful' kind. When we say that one human being has deceived another, we often imply that the deceiver knew what he was doing: President Nixon, we might say, wanted citizens to believe he was innocent, but knew he wasn't, so he behaved in a manner that would achieve certain goals. How would we prove that animals engaged in similar conscious, deceitful acts? How would we eliminate simpler explanations of the animals' behavior?

The first point to be made about any study of so-called 'conscious' behavior is that we should not set our expectations too high. If we aim for a rigorous, formal definition that offers a widely applicable characterization of different behaviors as 'conscious' or 'unconscious' we are likely to fail. Such failure would occur because, for the most part, we don't yet know what it is that we are attempting to define. On the other hand, if we aim for a working definition that may be formally incomplete but nevertheless focuses our attention on certain issues, this may in turn lead us to design better observations and experiments, and in the long run we are more likely to succeed in our ultimate goal.

Given this limited aim, Daniel Dennett (1978) has offered a philosophical analysis of some complex behavior in a form that may be useful to those studying animal communication, particularly 'deceitful' behavior. He begins by arguing that within the animal kingdom there is a continuum of what we might call mental states, from the simplest self-regulatory systems to our own mind. Within this continuum, Dennett argues, there are unlikely to be abrupt discontinuities. Thus our aim (see above) should not be to call some species 'conscious' and others not. Instead, we should set out to examine degrees of complexity by quantifying our own descriptions of animal behavior. How complex do *our* terms need to be to give the simplest satisfactory description of a particular observation or experiment?

Consider the case of a vervet monkey who gives a particular alarm call, thus warning its fellow group members that there is a leopard nearby (Seyfarth, this volume). What are the important questions to be asked, and how might we quantify the human terms needed to describe such behavior? One starting point (Seyfarth *et al.* 1982) might be to describe the monkey's behavior in the simplest terms possible: the monkey gives an alarm call because it is excited (what Dennett would call 'zero order intentionality'). We may discount this hypothesis because of the observation that lone monkeys who encounter predators do not give alarm calls (R. M. Seyfarth & D. L. Cheney unpublished). Of course, there are other possibilities, and no exhaustive hypothesis testing is implied here. I simply introduce the manner in which an explanation might proceed. An alternative hypothesis might argue that the monkey gives an alarm because it *wants* the other monkeys to run into a tree (first order intentionality). The observation that monkeys will alarm-call at leopards even if all the members of their group are in trees suggests that this explanation may also be eliminated. We may further hypothesize that the monkey gives an alarm call because it *wants* others to *believe* that there is either (*a*) something interesting nearby, (*b*) a predator nearby, or (*c*) a specific kind of predator nearby. All of these would be second order intentionality. To date, field observations suggest that explanations (*a*) and (*b*) can be eliminated, leaving explanation (*c*) as that most strongly supported by existing data. Throughout such an exercise, the main purpose of Dennett's scheme is to answer the question 'How complex must *our human terms* be to describe a particular pattern of behavior? When we compare different sorts of animal communication, which require the most complex terms?'

As a second example, consider the experiments of Woodruff & Premack (1979), reviewed above by Quiatt. Here one question might be: why isn't the chimpanzee puzzled when the 'bad' trainer, having gone to the incorrect box, fails to go to the correct box? The chimp's lack of puzzlement (if it exists) suggests that he either (*a*) doesn't really understand the relation between the trainer's knowledge and his actions, or (*b*) believes that the trainer is exceedingly stupid.

On the basis of Premack & Woodruff's evidence, can we say that the chimpanzee *wanted* the trainer to *believe* that the food was in a particular container? Here Dennett & E. S. Savage-Rumbaugh (quoted in Seyfarth *et al.* 1982) suggest an intriguing experiment: place the food in plastic, see-through containers, and instruct the 'bad' trainer to look at these containers as he enters the room. If the chimpanzee can instantly recognize a logical inconsistency between (*a*) what it wants the trainer to believe, and (*b*) what the trainer obviously knows, then we would clearly need more complex terms to describe the animal's behavior.

At this point it should be emphasized that Dennett's analytical scheme deals with the language we humans need to describe animal behavior. Such descriptions are definitely not the same thing as the animal's 'thoughts', 'wishes', or 'desires'. For example, suppose we were to conduct a series of experiments that proved, to our satisfaction, that the following was the simplest explanation of a vervet monkey's alarm-calling behavior: 'the monkey alarms because it wants others to know that there is a leopard nearby'. This doesn't mean that the above statement is true. Instead, it means that the monkey's behavior is sufficiently complex that we are forced to use the above human terms to describe it. We have taken a small step in quantifying our descriptions of behavior – but we should not mistake such descriptions for explanations.

Nevertheless, just because such limitations exist, we should not be blind to the practical advantages of Dennett's scheme. If used with care, it allows us to be more precise when we make comparisons between different sorts of animal communication, and it promises to suggest new observations and experiments. It is not a panacea, but it does provide temporary help as we grope for an explanation to some extremely complex phenomena.

References

Dennett, D. (1978). *Brainstorms. Philosophical Essays on Mind and Psychology.* Montgomery, Vermont: Bradford Books.

Seyfarth, R. M. *et al.* (1982). Communication as evidence of thinking. In *Animal Mind–Human Mind,* ed. D. R. Griffin. Berlin: Dahlem Konferenzen, Springer Verlag.

Woodruff, G. & Premack, D. (1979). Intentional communication in the chimpanzee: the development of deception. *Cognition,* **7**, 333–62.

B. Primate communication systems in use

2. What the vocalizations of monkeys mean to humans and what they mean to the monkeys themselves

ROBERT M. SEYFARTH

2.1. Introduction

When one person uses a word in speaking to another, three important events take place. First, the speaker creates a sound with his vocal organs. For all those who share the speaker's language, this sound represents, or 'stands for', a relatively specific set of information. Second, the speaker has some mental image of an object or a concept, as well as the intent to communicate this image to another. Third, the listener assimilates the information conveyed by the sound, and responds to it. In most cases this response is imperceptible, not only because most words do not evoke extreme behavioral changes but also because it is difficult to monitor information processing in the brain.

These three characteristics of language – by no means exhaustive – highlight the problems we face when we attempt to compare the signals of monkeys and apes with human words. For example, consider the theoretical framework that has long dominated research on the natural signals of non-human primates. While we readily assume that the noises made by humans can stand for some object or concept, ethologists working with non-human primates have generally begun their work with the quite different assumption that vocalizations are simply a manifestation of some internal, emotional state. Rather than conveying specific information about objects or events in the external world, animal signals are assumed to specify only that an individual is, say, nervous, hungry, or excited. Given these fundamentally different assumptions, it is not surprising that scientists have uncovered few similarities between the way humans use words and the way monkeys and apes use vocalizations.

A second problem for comparative research derives from our inability to assess the meaning of non-human primate signals. As Marler (1961) originally pointed out, the only way to measure the 'meaning' of an animal signal is to measure the responses it evokes in others. Obviously, the use of such a technique for the analysis of human words would yield few insights. Nevertheless, this crude technique is the one to which human observers are limited when they analyze the sounds of non-human primates, even though it invariably underestimates both the variety and the specificity of meaning in non-human primate signals.

Finally, it is virtually impossible to measure the cognitive processes or intentions involved in non-human primate signals (cf. Dennett 1980; Griffin 1981). We feel relatively safe in assuming that humans intend to communicate certain information when they speak, and we generally assume that such intention is somehow unique to human language. Consequently, our explanations of the motives behind human speech are generally phrased in cognitive terms (e.g. 'The man shouted "Fire!" because he wanted to communicate certain danger to others'). However, when we analyze animal vocalizations – where no data on intention are immediately apparent – we carefully avoid the issue of cognition (e.g. 'The monkey gave an alarm call because it was afraid'). Here again, methodological limitations have led us to adopt a theoretical dichotomy that may not, in fact, exist.

It is clear that in our current attempts to compare the signals of monkeys, apes, and humans some questions are more easily answered than others. Because we cannot usually measure the cognitive processes of a vocalizing primate, our interpretation of the meaning of its vocalization must be inferred through the responses such vocalizations evoke in others (but see Cheney, this volume). Data on responses do, however, permit us to examine the function of animal signals: how primates perceive their vocalizations, and how they use vocalizations to inform or deceive their fellow group members.

In the remainder of this chapter I discuss how field experiments can be used to study the function of vocalizations in East African vervet monkeys (*Cercopithecus aethiops*). I begin by reviewing briefly some of the work that has been conducted on the alarm calls given by vervet monkeys to different species of predator. My primary focus, however, will be on the grunts used by vervet monkeys during social interactions with each other. I argue that, regardless of their precise meaning or causation, the vervets' grunts share many functional similarities with human words.

2.2. Subjects and study area

The research described here (and in Cheney, this volume) has been carried out over a number of years in Amboseli National Park, which lies on the southern border of Kenya at the foot of Mt Kilimanjaro. Within the park, the vervet monkeys under study inhabit an area of semi-arid acacia savanna near two permanent waterholes. Their social organization consists of a number of discrete groups, ranging in size from 10 to 30 individuals. A typical group contains a number of adult males, adult females, and their offspring. Each group occupies a small territory, which is defended against incursions from other groups (Cheney 1981; Cheney & Seyfarth 1982*b*).

Female vervet monkeys generally remain in the group where they were born throughout their lives, while males transfer from one group to another, beginning at around sexual maturity. Within each group, individuals can be ranked in a linear dominance hierarchy, which determines priority of access to food, social partners, or other limited resources (Seyfarth 1980; Wrangham 1981). Immature vervets acquire ranks immediately below those of their mothers (Cheney, 1983).

2.3. Previous research: Vervet monkey alarm calls

Field research on the alarm calls given by vervet monkeys to different predators has indicated that the vocalizations of non-human primates can effectively function as if to designate objects in the external world. Vervet monkeys give acoustically different alarm calls to each of their four main predators: leopards, eagles, pythons, and baboons (Struhsaker 1967). Playback experiments using tape-recordings of three of these alarm calls have shown that, even in the absence of actual predators, each alarm call type evokes a qualitatively different set of responses from the monkeys (Seyfarth, Cheney & Marler 1980*a*,*b*). When the monkeys are foraging on the ground, for example, playback of an alarm call originally given to a leopard causes them to run into trees. Playback of an eagle alarm causes them to look up or run into bushes, and playback of a snake alarm causes them to look down, scanning the ground around them. Moreover, variation in the length and amplitude of different calls (two measures one might associate with variation in caller 'arousal') does not blur the distinction among different response types. Variation in the acoustic properties of different calls seems to be the only measure both necessary and sufficient to evoke observed differences in response.

Because playback of the vervets' alarm calls in the absence of predators evoked the same response as would the predators themselves, and because responses were largely independent of caller arousal, the calls may be regarded as a rudimentary form of representational signalling, in which relatively arbitrary, non-iconic signs are used to designate particular referents external to the signaller (see also Hockett 1960; Altmann 1967). The prevalence throughout the animal kingdom of similar predator-class or predator-specific alarm calls (e.g. Melchior 1971; Struhsaker 1975; Ryden 1978) suggests that many animal signals may refer to or 'stand for' events external to the communicator.

Experiments on alarm calls thus argue against the traditional dichotomy separating the referential signals of language from the 'affective' signals of non-linguistic species. In many primates, however, alarm calls represent only a small subset of the animals' vocal repertoire, and it might be argued that the referential specificity of alarm calls is not representative of the monkeys' entire communicative system. Most vocalizations by monkeys and apes consist of grunts, screams, and chutters, uttered during social interactions with other group members. Do such calls also function as though they designate events in the external world?

2.4. Vervet monkey grunts

When one is following vervet monkeys during a normal day's observation, the most common vocalization one hears is a low-pitched, pulsatile grunt, originally described as a 'progression grunt' or 'woof' by Struhsaker (1967). As Struhsaker noted, this grunt is given in a variety of social contexts. For example, a monkey may grunt as it approaches an individual who is dominant to itself, or as it approaches a subordinate. In some cases, a monkey may grunt as it watches another animal initiate a group movement across an open plain. A grunt may also be given when a monkey has apparently just spotted the members of another group. Even to an experienced human listener, there are no immediately obvious audible differences among grunts, either from one context to another or across individuals (Figure 2.1; see also below). Although the calls may occasionally be answered by grunts from another monkey, in most cases the grunts given in each of the contexts described above evoke no salient behavioral response. Changes in the direction of gaze, which are difficult to measure under observational conditions, appear to be the only obvious response to grunts.

Vervet monkey grunts highlight the problems we face when we attempt to understand the meaning of primate signals. Watching

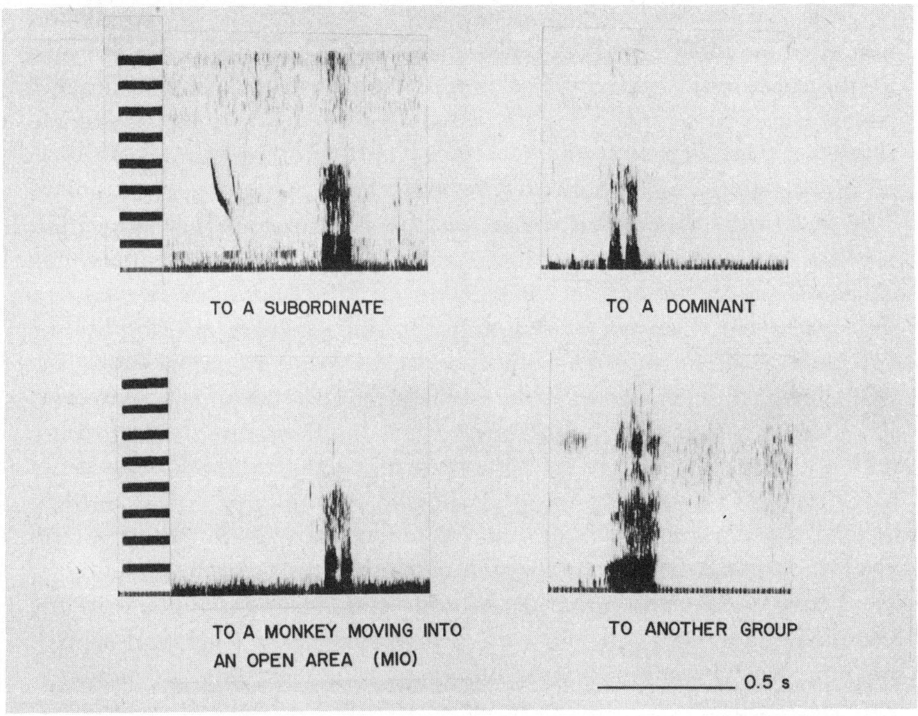

Figure 2.1. Sonograms of representative grunts given by one adult female in four different social contexts. The *x*-axis shows time and the *y*-axis shows frequency in units of 1 kHz.

monkeys grunt to each other is in many respects like watching humans engaged in conversation, without being able to hear what is being said. Our only way to assess the meaning of different grunts is from the responses of those nearby – and these responses, for both the monkeys and conversing humans, may be infrequent, subtle, or entirely undetectable.

Faced with this methodological dilemma, scientists have traditionally concluded that vervet monkeys, like other non-human primates, are using one vocalization in a variety of different contexts. Each grunt, they would argue, is a manifestation of a particular level of arousal, providing information about the vocalizer's motivation, identity, or perhaps serving as a kind of 'contact call'. Given the rather general nature of the information they convey, any variation in the responses evoked by different grunts can be accounted for by variation in the contexts in which they are given.

In contrast, both Green's (1975) work on Japanese macaques and our own earlier work on vervet alarm calls suggest an alternative hypothesis: namely, that what appears to a human listener to be one grunt is in fact a number of different grunts. Each grunt type, in and of itself, conveys specific information that depends more on the acoustic properties of each call than on the context in which it is given. In 1980, we conducted a series of field playback experiments to test between these two hypotheses.

In designing these experiments, our general approach was as follows (for further details see Cheney & Seyfarth 1982*a*). First, grunts from the same individual were tape-recorded in each of the four different contexts described above. Second, over a number of months each of these grunts was played back to subjects from a concealed speaker, and the subjects' responses to the calls were filmed. For example, a given subject would be played the 'grunt to a dominant' on one day, and then, three or more days later, the same subject would be played the 'grunt to another group'.

Throughout our experiments, we allowed social context to vary freely. Trials were conducted, for example, when there were dominant or subordinate animals nearby, when the group was foraging or resting, when they were at the center or the edge of their range, and so on. We reasoned that, if the grunts were really only one vocalization whose meaning was largely determined by context, subjects should show no consistent differences in responding to the calls. Instead, responses to playback should be a function of the variable context within which calls were presented. On the other hand, if each of the grunt types was different, and if each carried a relatively specific meaning, we should expect consistent differences in the response to each grunt type regardless of the varying circumstances in which they were played. Our overall method thus involved a series of paired playback experiments in which subjects were asked 'Is grunt A different from grunt B? If so, is grunt A different from grunt C? Do grunts B and C differ?', and so on.

When conducting an experiment, we first decided which grunt was to be played, and which individual would serve as a subject. Anticipating the movement of the subject, we then selected a likely location, hid a speaker in a bush or tall grass, and prepared to film the subject using a sound movie camera. If the subject came to within three metres of the speaker, and if all of our other pre-trial conditions had been satisfied (Cheney & Seyfarth 1982*a*), the trial began. As noted earlier, a given subject heard first one grunt type and then, at least three days later, a second grunt that had originally been recorded in a different social context. The order in which grunt types were presented was systematically varied, so that no grunt type consistently appeared as the first or second stimulus in any series of paired trials. Throughout all trials, we attempted to ensure that our experimental protocol mimicked natural conditions as much as possible. Vervets frequently forage in dense bush near waterholes, or in shrubs and vines at the base of trees, and they often grunt when they are apparently out of sight of one another. We therefore felt justified in assuming that a grunt played from a speaker, as if uttered by a monkey behind a bush or tree, was not anomalous.

A playback experiment was possible on only 10–20 per cent of all occasions when we were prepared to conduct a trial. Attempts to conduct a trial were frustrated when, for example, the predesignated subject moved in an unanticipated direction, away from our chosen location, or when the subject concealed itself in dense bush. Alternatively, the subject situated itself satisfactorily, but the individual whose grunt was about to be played emerged in full view. Clearly, under these conditions we could not conduct a realistic experiment, so we had to wait until the animal whose call was being played was, to the best of our knowledge, out of sight of the subject.

In each trial, the subject was filmed for ten seconds before and ten seconds after a grunt was played. When analyzing the films of our experiments, the duration of response was defined as the duration of a given behavior after playback minus the duration of the same behavior before playback. This allowed us to control for the subject's behavior in the second immediately preceding the grunt playback. If playback increased a given behavior, the duration of response would be positive; if playback decreased a given behavior the duration of response would be negative; and if playback produced no change the duration of response would be zero.

As an example of the results we obtained, I have chosen one experimental series, in which 18 subjects were asked to compare a grunt originally given to a dominant and a grunt originally given to another group. In this comparison, playbacks evoked two sorts of response: subjects either looked toward the speaker for longer after playback than before, or they looked out, orienting their face in the direction the speaker was pointed. There were consistent differences in the responses evoked by the two calls. Grunts to a dominant generally caused subjects to look toward the speaker, while grunts to another group caused subjects to look in the direction the speaker was pointed (Figure 2.2). In other words, grunts to another group seemed to have the function of directing the listener's attention away from the speaker, and toward a place where, under natural conditions, the vocalizer would have been facing.

The results of all our experiments are summarized in Table 2.1, which gives, for each paired comparison, the behavioral measure or measures

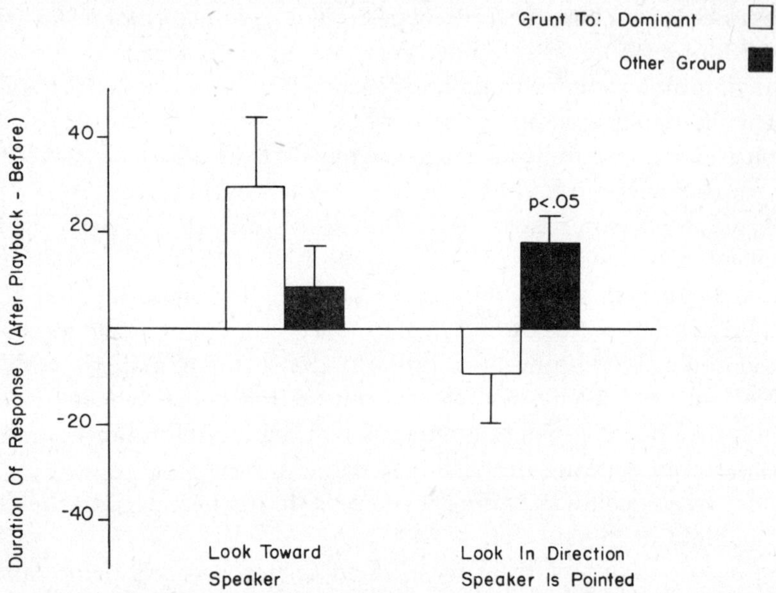

Figure 2.2. Duration of responses to playback of the grunt to a dominant and the grunt to another group. The *y*-axis shows duration of response (defined in text) measured in frames of film at 18 frames s^{-1}. Histograms show mean and standard errors for 18 subjects.

that showed a significant difference between the response to one grunt as compared with the response to another. It should be noted that the grunt to a subordinate was the only vocalization that caused animals to move away from the speaker. Thus even though the grunt to a subordinate was not tested directly against the grunt to an animal as it moves into an open area or the grunt to another group (Table 2.1), we have indirect evidence that the animals would also make these discriminations.

In other words, each grunt type evoked a different response. By their behavior, the monkeys seem to be telling us that, although grunts sound more or less the same to us, to the monkeys themselves each grunt type transmits a different and rather specific set of information. Moreover, this information can in some cases act as though it refers to events external to the signalling individual, such as the approach of another group or the movement of another individual into an open area. And finally, the monkeys responded differently to different grunts even though our experiments were carried out in a variety of social circumstances. This suggests that the meaning of each grunt depends more on the acoustic properties of the call itself than on the context in which it is given (see also below).

Table 2.1. *A summary of significant differences between the responses evoked by differing grunts in paired playback trials*[a]

	Grunt given to:			
	Subordinate	Dominant	MIO	Other group
Subordinate		1	(Not tested)	
Dominant			2	4
MIO				3, 5

[a]Measures tested were: 1, move away from speaker; 2, duration of looking toward speaker; 3, latency to look toward speaker; 4, duration of looking in direction speaker pointed; 5, latency to look in direction speaker pointed. Entries in each cell indicate a significant difference at the 0.05 level (two-tailed), except for measures 3 and 5, which differed at the 0.07 level (two-tailed).

2.5. Acoustic cues

Our experiments indicate that what humans initially perceive as one grunt the monkeys perceive as at least four. At the same time, however, there are no immediately obvious acoustic distinctions that allow a human listener to distinguish among the various grunt types. We have therefore begun a program of laboratory analysis to determine, first, what acoustic cues the monkeys might use to distinguish between different grunts, and second, whether monkeys in different groups (or in different areas of East Africa) use different or similar acoustic 'conventions' to tell one grunt from another.

As a first step in our analysis, we examined grunts given by one adult female (TD) in the four social contexts described above: when approaching a dominant, when approaching a subordinate, when watching another monkey move into an open area, and when looking at another group. For each of 16 acoustic measures, an overall statistical test was used to determine the degree of heterogeneity across grunt types. If an acoustic feature showed significant heterogeneity, we concluded that the monkeys might potentially make use of this feature when distinguishing between TD's grunts. We then repeated this analysis using grunts from five other adult females, representing a total of three social groups whose contiguous ranges covered an area of roughly 1.5 square kilometres.

The grunts of adult female TD, as well as those of other vervet monkeys, appeared on a frequency v. time display as rather noisy, non-tonal sounds (Figure 2.1; see also Struhsaker 1967). Nevertheless, detailed spectral analysis revealed a number of acoustic measures that could potentially allow a listener to distinguish between grunts given in different contexts. TD's grunt to a dominant, for example, typically contained one frequency peak, usually around 310 Hz. Grunts to a subordinate, grunts to an animal moving into an open area, and grunts to another group contained a second peak, usually between 690 and 1500 Hz. In grunts to a subordinate the two peaks were of roughly equal amplitude. In MIO grunts the lower peak was stronger in amplitude, while in grunts to another group the upper frequency peak was stronger in amplitude (Fig. 2.3). Each of these measures showed significant heterogeneity across grunt types (Cheney & Seyfarth 1982*a*), suggesting that they could have been used to distinguish between TD's grunts.

Results also indicated considerable 'agreement' among adult females in the acoustic cues that distinguished between different grunt types, even though these females came from different social groups. Of the 16 acoustic measures tested across six adult females, seven measures were non-significant for all subjects, and three other measures were significant for only one subject. In contrast, two measures were significant for four of the six subjects, and three measures were significant for five subjects.

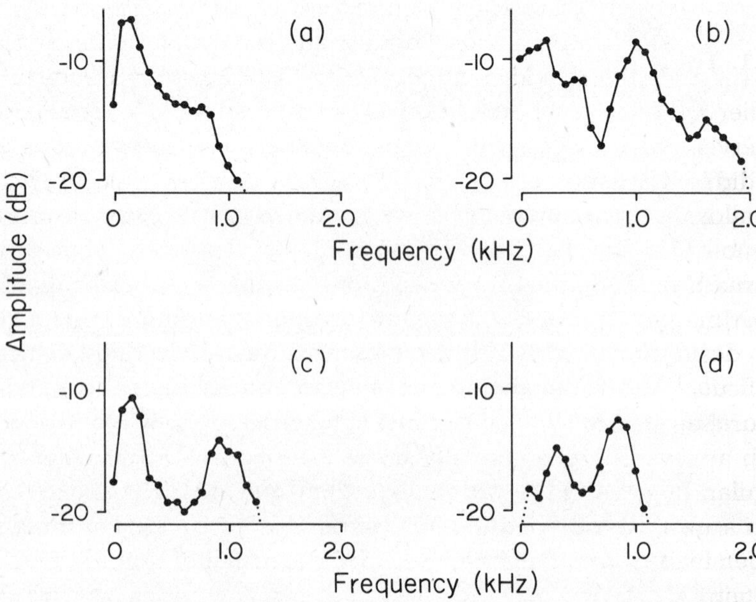

Figure 2.3. Power spectra of four grunts given by the same adult female in four different social contexts: (a) when approaching a dominant, (b) when approaching a subordinate, (c) when watching another animal move into an open area, and (d) when looking at another group.

In other words, it seems likely that all individuals within the area where our work is conducted make use of the same acoustic distinctions when giving grunts in different contexts. Our analysis of acoustic cues has only begun, however, and it would be premature to rule out the possibility that non-human primates, like many songbirds, have local dialects that differ across relatively small geographic areas, and which are learned during ontogeny (Green & Marler 1979).

2.6. Discussion

The experiments described above raise a number of general issues dealing with the nature of vocal signals in human and non-human primates. They indicate, for example, that the information transmitted by vervet monkey grunts is quite specific, and that in many cases the grunts function in a manner that serves to designate events external to the vocalizer, such as another monkey moving into an open area or the approach of another group. At the same time, however, I think it is important to point out the limitations of our experiments, and to emphasize that they tell us relatively little about the *precise* meaning of different grunts or the causal mechanisms that underlie them.

2.6.1. The problem in assessing meaning

Earlier I argued that the responses evoked by a vocalization are the observer's only guide to its meaning. In our experiments on vervet grunts we did indeed evoke different responses to different signals, but this did not allow us to determine the precise meaning of a given signal. The grunt to another group could designate 'There is another group approaching'; alternatively, it could simply mean 'Look over there'. Similarly, alarm call experiments did not permit us to determine whether the leopard alarm call designated, for example, a particular species of mammal or a particular kind of danger. Responses in playback experiments (or under natural conditions) are not equivalent to meaning, they simply provide us with a means of determining whether monkeys regard two signals as 'similar' or 'different'.

In a more general sense, however, data on responses can help us to understand the array of possible meanings that an animal signal *might* contain. Response data may tell us, for instance, that signals can potentially relate to external events. Study of such referential signals can, in turn, increase our understanding of meaning by revealing how animals classify objects for the apparent purpose of signalling about them. We can learn, for example, that monkeys distinguish terrestrial predators from raptors, because they use the same vocalization when signalling about objects within a class and different vocalizations when signalling about objects within different classes (Cheney, this volume). Granted, this is not comprehension of meaning at the level we are used to in humans, but it is, perhaps, a step in the right direction.

Finally, it is clear that we run the risk of over-simplification if we restrict our comparisons to signals which we describe as either 'affective' or 'semantic'. As I mentioned earlier, when we started our research in 1977 there was a well-established dichotomy distinguishing the so-called emotional signals of monkeys and apes from the more referential words used in human language. At the time, such a dichotomy created a useful starting point, because it prompted us to ask whether primate signals might, in some way or another, carry any information that could be called referential. Now that we are in the midst of our research , however, this dichotomous approach is much less useful, since it involves an over-simplification of both human and non-human primate communication. The vocalizations of monkeys and apes can obviously convey information *both* about the signaller's emotional state and about events going on elsewhere. The precise balance between these two components of meaning in any given situation remains to be determined.

2.6.2. Parallels with human words

Because vervet monkey grunts transmit quite specific information, and because in some cases they clearly relate to external events, it is tempting to compare them with human words, where individuals often have a mental representation of some object or event they wish to communicate. Clearly, however, such a comparison is premature, since our experiments can only measure the responses evoked by grunts, and not what is going on in the mind of the vocalizer.

Nevertheless, regardless of their precise causation, vervet grunts can potentially *function* in a manner analogous to words because of the nature of the information they convey to others. When a monkey hears a grunt, for example, he is immediately informed of many of the fine details of the social behavior going on, even though he may be out of sight of the vocalizer, and even though the vocalizer himself may not be involved. At this point, the monkey's system of communication, whatever its physiological or cognitive basis, begins to take on many of the *functional* properties of a system of communication based on true representation.

Although the experiments described here give us a somewhat better understanding of how vervet monkey grunts function, they do not elucidate exactly what that function is, nor do they explain why natural selection has given rise to so many different signals. While it is not difficult to imagine the advantage of developing different alarm calls for predators with different hunting strategies, the benefits of such representational signalling within the group are less clear. Of course, once this behavior has arisen it is easy to imagine how such unambiguous and precise communication might benefit the individuals involved. It is considerably more difficult, however, to imagine how natural selection first caused these signals to appear. As is often the case, one must distinguish the evolutionary pressures that first cause a behavior to appear from those that cause it to persist once it has reached a certain frequency.

Thinking of grunts in terms of their function reminds us that any animal species, including humans, will only develop elaborate, complex signals when natural selection makes it necessary for them to do so (e.g. Humphrey 1976). Our research thus far has dealt with a relatively simple subset of grunts, namely those given in four unambiguous social contexts. Under normal conditions, however, vervet monkeys often grunt in circumstances that are multi-contextual: for example, when a vocalizer is both approaching a dominant and watching another monkey move into an open area. Such conditions would seem to place strong selective pressure on individuals to use a number of different grunts in sequence (perhaps distinguishing among different grunt orders), or to alter the acoustic properties of grunts in a manner that produces new messages. In short, complex social situations may create selection pressures for the evolution of syntax. Whether monkeys respond to such situations in a manner that suggests syntactical abilities will be one subject for future research.

Acknowledgements

I thank D. Ploog and K. Scherer for helpful comments on the manuscript. Research was supported by the H. F. Guggenheim Foundation and the National Science Foundation.

References

Altmann, S. A. (1967). The structure of primate social communication. In *Social Communication Among Primates*, ed. S. A. Altmann, pp. 325–62. Chicago: University of Chicago Press.

Cheney, D. L. (1981). Inter-group encounters among free-ranging vervet monkeys. *Folia Primatologica,* **35**, 124–46.

Cheney, D. L. (1983). Extra-familial alliances among vervet monkeys. In *Relationships and Social Structure in Some Non-Human Primates,* ed. R. A. Hinde. Cambridge University Press.

Cheney, D. L. & Seyfarth, R. M. (1982*a*). How vervet monkeys perceive their grunts: field playback experiments. *Animal Behaviour,* **30**, 739–51.

Cheney, D. L. & Seyfarth, R. M. (1982*b*). Recognition of individuals within and between groups of free-ranging primates. *American Zoologist,* **22**, 519–29.

Dennett, D. (1978). *Brianstorms, Philosophical Essays on Mind and Psychology.* Montgomery, Vt.: Bradford Books.

Green, S. (1975). Communication by a graded vocal system in Japanese monkeys. In *Primate Behavior,* vol. 4, ed. L. A. Rosenblum, pp. 1–102. New York: Academic Press.

Green, S. & Marler, P. (1979). The analysis of animal communication. In *Handbook of Behavioral Neurobiology,* vol. 3: *Social Behavior,* ed. P. Marler & J. G. Vandenbergh, pp. 73–158. New York: Plenum Press.

Griffin, D. R. (1981). *The Question of Animal Awareness,* 2nd edn. New York: Rockefeller University Press.

Hockett, C. F. (1960). Logical considerations in the study of animal communication. In *Animal Sounds and Communication,* ed. W. E. Lanyon & W. L. Tavolga, pp. 292–340. Washington: American Institute of Biological Sciences.

Humphrey, N. K. (1976). The social function of intellect. In *Growing Points in Ethology,* ed. P. P. G. Bateson & R. A. Hinde, pp. 303–18. Cambridge University Press.

Marler, P. (1961). The logical analysis of animal communication. *Journal of Theoretical Biology,* **1**, 295–317.

Melchior, H. R. (1971). Characteristics of arctic ground squirrel alarm calls. *Oecologia,* **7**, 184–90.

Ryden, O. (1978). Differential responsiveness of great tit nestlings, *Parus major,* to natural auditory stimuli. *Zeitschrift für Tierpsychologie,* **47**, 236–53.

Seyfarth, R. M. (1980). The distribution of grooming and related behaviors in adult female vervet monkeys. *Animal Behaviour,* **28**, 793–813.

Seyfarth, R. M., Cheney, D. L. & Marler, P. (1980*a*). Vervet monkey alarm calls: semantic communication in a free-ranging primate. *Animal Behaviour,* **28**, 1070–94.

Seyfarth, R. M., Cheney, D. L. & Marler, P. (1980*b*). Monkey responses to three different alarm calls: evidence for semantic communication and predator classification. *Science,* **210**, 801–3.

Struhsaker, T. T. (1967). Auditory communication among vervet monkeys (*Cercopithecus aethiops*). In *Social Communication Among Primates,* ed. S. A. Altmann, pp. 281–324. Chicago: University of Chicago Press.

Struhsaker, T. T. (1975). *The Red Colobus Monkey.* Chicago: University of Chicago Press.

Wrangham, R. W. (1981). Drinking competition among vervet monkeys. *Animal Behaviour,* **29**, 904–10.

COMMENT

D. PLOOG

Seyfarth's approach to get at the meaning of vocal signals seems to me a very promising one. I'd like to comment on the one remark he made on habituation to the leopard warning call and to the other warning calls. In the early 1960s, when we worked on the vocal repertoire of the squirrel monkey, we found two alarm calls: firstly a very short peep for aerial predators (elicited by moving objects of certain size and velocity across a glass dome over the cage) and secondly a yapping call, repeated by the whole group (elicited by slow-moving strange objects on the ground, preferably furry dummies). In the first case the monkeys stay put and remain absolutely silent for several seconds; in the second case they move quickly up to a higher place as far off the ground as possible, orient towards the 'ground predator' and yap with increasing intensity, comparable to the mobbing response of songbirds. During playback experiments we found that both calls habituate rather soon. If (in the case of yapping) the strange object remains in sight, the monkeys begin to explore the object by approaching it, dashing away, approaching again and eventually touching it. I conclude that, by habituating quickly to the leopard warning call, the vervets assess the situation and thus do not respond repeatedly to a leopard which is obviously not present.

References

Ploog, D. (1967). The behavior of squirrel monkeys (*Saimiri sciureus*) as revealed by sociometry, bioacoustics, and brain stimulation. In *Social Communication among Primates,* ed. S. A. Altmann, pp. 149–84. Chicago: University of Chicago Press.

Winter, P., Ploog, D. & Latta, J. (1966). Vocal repertoire of the squirrel monkey (*Saimiri sciureus*), its analysis and significances. *Experimental Brain Research,* **1**, 359–84.

3. Category formation in vervet monkeys

DOROTHY L. CHENEY

Playback experiments of both the predator alarm calls of vervet monkeys and the grunts used by vervets during social interactions have indicated that non-human primate vocalization may effectively function to designate objects or events in the external world (Cheney & Seyfarth 1982*a*; Seyfarth, Cheney & Marler 1980; Seyfarth this volume). These data reveal much about the ways monkeys communicate with each other, and about the possible selective pressures that might have favored the evolution of representational signalling in human and non-human primates. Perhaps even more importantly, however, they provide the human observer with the intriguing opportunity to investigate how monkeys classify objects and events in the world around them. At present, our investigations of monkey vocalizations are limited, in the sense that they can examine only the responses that calls evoke in listeners, and not the vocalizers' physiological or affective states (Seyfarth this volume). However, whenever an organism communicates 'about' objects external to itself, it effectively classifies objects in its environment, and in so doing reveals some of the cognitive processes that underlie its signals.

In this chapter, I investigate the manner in which vervet monkeys perceive and classify objects and events in the external world. In the first part of the chapter, I examine whether the association that vervets make between a particular acoustic unit and its referent is arbitrary or iconic, and discuss the extent to which vervets may be said to classify objects as graded continua or as discrete and separate entities. In the second part of the chapter, I use data on alarm calls and other vocalizations to argue that classifications made by vervets both of one another and of other objects are complex and hierarchical, in many respects resembling the multilevel taxonomies created by humans to describe their own social organizations.

3.1. Study subjects

The vervet monkeys described in this paper have been observed continuously since at least 1977 in Amboseli National Park, Kenya. Three social groups inhabiting contiguous home ranges have been studied intensively, and an additional eight neighboring groups have been regularly censused. The group range in size from 11 to 25 animals, and typically consist of a number of adult males, adult females, and their immature offspring. Females remain in their natal groups throughout their lives, while males transfer to neighboring groups at around sexual maturity. Within the group, both adult males and adult females can be ranked in linear dominance hierarchies, based on the direction of approach–retreat interactions. Juveniles assume ranks similar to those of their mothers during aggressive disputes, such that all the members of a given family share similar ranks (Cheney 1983; see also Seyfarth this volume).

Individuals in each of the three study groups are sampled on a regular basis. During such sampling sessions, all vocalizations exchanged among individuals are noted, and as many vocalizations as possible are tape-recorded. In addition, all alarm calls are noted on an *ad libitum* basis, as well as the responses of nearby animals (see Seyfarth *et al.* 1980).

3.2. Part one

Playback experiments using both vervet monkey alarm calls and grunts have indicated that the monkeys' calls may function to designate objects or events in the external world (Seyfarth this volume). How arbitrary, however, is the association between a particular acoustic unit and its referent? Investigations of the signalling abilities of captive apes have often focused on the extent to which the signs learned by apes are arbitrary or iconic with respect to the objects they represent (Ristau & Robbins 1979). The question is a crucial one, because it is often assumed that one feature that distinguishes human language from animal communication is the ability of humans to make relatively arbitrary associations between particular acoustic units and their referents.

Our ability to assign arbitrary and similar-sounding acoustic units to very different referents presupposes an ability to classify objects into discrete categories. In contrast, it is often implicitly assumed that the graded, apparently iconic communication of animals demonstrates a similarly graded and continuous perception of objects or events. However, recent evidence that non-human primates use similar-sounding acoustic units to designate very different objects appears to contradict this assumption (Green 1975; Cheney & Seyfarth 1982*a*). These data suggest that animals may perceive objects as discrete and separate classes, and that the association between a particular acoustic unit and its referent may often be more arbitrary than previously imagined.

Human words often consist of homologous acoustic units which nevertheless have very different referents. The words 'bear' and 'pear', for example, refer to very different and discrete objects, despite the fact that the two words share many acoustic similarities. Similarly, homologous referents may be represented by very different acoustic units, as, for example, 'bearish' and 'ursine'. Two features of human language permit the creation of words of these types. Firstly, as mentioned previously, the association between a given acoustic unit and its referent is relatively arbitrary; words do not necessarily sound like the objects they designate. Secondly, when presented with a graded continuum of acoustic units (ranging, for example, from 'bear' at one extreme to 'pear' at the other), humans often perceive the sounds categorically (Liberman *et al.* 1967). To what extent do the calls of non-human primates share these features of human language?

3.2.1. The relation between acoustically similar sounds and their referents

Laboratory experiments have demonstrated that Japanese macaques are capable of perceiving a graded continuum of their own calls in a categorical manner (Zoloth *et al.* 1979). The extent to which similar categorical perception of species-specific calls exists in the wild is not known, since playback experiments to test this hypothesis have not been conducted on free-ranging monkeys. Nevertheless, it seems reasonable to assume that, under natural conditions, Japanese macaques also perceive their calls in a categorical manner. Moreover, there is indirect evidence of categorical perception of species-specific calls in free-ranging vervet monkeys. Playback of similar-sounding grunts, for example, evokes consistently different responses from vervets, even across a variety of contexts (Cheney & Seyfarth 1982*a*; Seyfarth this volume).

Similarly, observational data from other types of vervet vocalizations indicate not only that similar-sounding acoustic units are often perceived as distinct and separate categories, but also that they may refer to very different objects. For example, the vervet monkeys' alarm-call to a python or poisonous snake is a rasping call (hereafter called a 'chutter') consisting of a series of short, relatively widely separated acoustic units. The energy

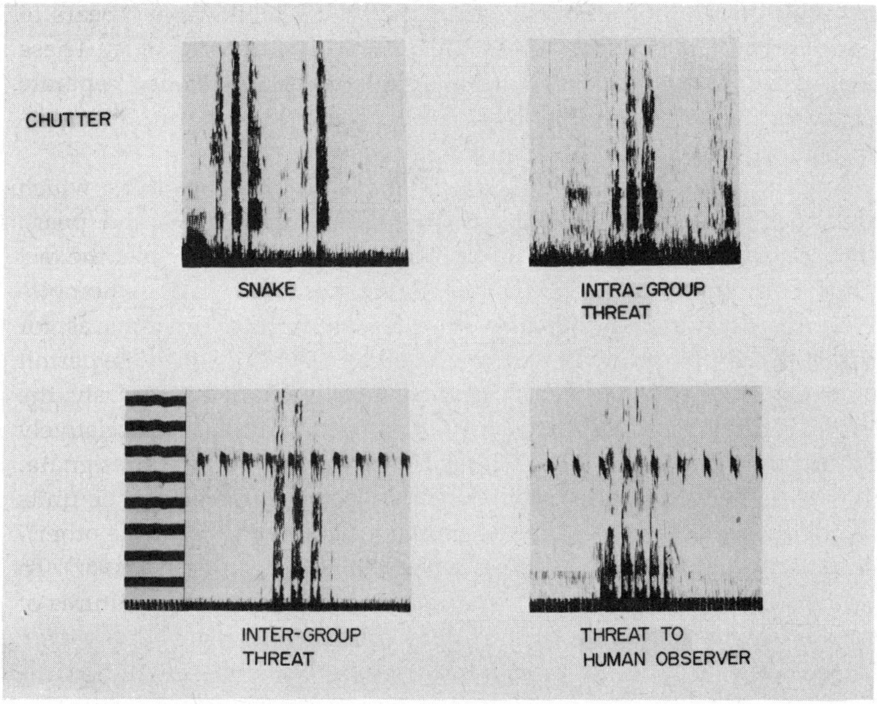

Figure 3.1. Sonograms of chutters uttered by vervet monkeys in four different contexts. The x-axis indicates time; y-axis indicates frequency in units of 1 kHz.

in these units is concentrated across a wide frequency band, which usually ranges from 0 to 16 kHz (Figure 3.1; see also Struhsaker 1967; Seyfarth *et al.* 1980). Calls with these acoustic properties are not, however, restricted solely to snake alarm calls. Similar-sounding chutters are used in three other, very different, contexts: when threatening a member of one's own group, when threatening a member of another group, and when threatening a human observer. While chutters uttered in each of these four contexts sound very similar to the human ear, there are small but consistent acoustic differences across different types of chutter (Struhsaker 1967; Seyfarth *et al.* 1980). In other words, although the calls share similar general acoustic properties, there are nevertheless consistent, significant differences that divide the calls into sub-types. These, in turn, are used in widely different social situations (Figure 3.1).

3.2.2. The relation between similar referents and acoustic units

There is indirect evidence, then, that vervet monkeys perceive similar-sounding acoustic units categorically, and that similar acoustic units may refer to different objects in the external world. Are there also instances, however, when very different acoustic units refer to similar objects? Again, field data on vervet monkeys provide indirect evidence of this, and offer further evidence of a relatively arbitrary association between a given acoustic unit and its referent.

As mentioned previously, vervet monkeys not only chutter when alarm calling at snakes, but also chutter when threatening a member of their own group, a member of another group, or a human observer. Chutters are not, however, the only vocalizations used in these contexts. When vervets threaten a member of their own group, they may either chutter, grunt, or pant-threat (Figure 3.2; see also Struhsaker 1967). Similarly, when vervets encounter members of another group, they either chutter, grunt, or utter a prolonged 'aarr' (Figure 3.2; see also Struhsaker 1967; Cheney & Seyfarth 1982*b*; Seyfarth this volume). When inter-group encounters escalate into aggressive fights, vervets often utter pant-threats and barks as well as chutters. Finally, the chutter is not the only vocalization used by vervets when they encounter other species of primates. Although sample sizes of the calls are small, vervets also appear to give acoustically distinct alarm calls to two species of potentially threatening primates: baboons and strange humans (usually Maasai tribesmen) (Figure 3.2). Alarm calls to each of these two primate species are acoustically distinct from alarm calls given to leopards, eagles, and snakes (Figure 3.2).

Within a given broad contextual category, therefore, vervet vocalizations do not necessarily share the same general acoustic properties. Vocalizations used during interactions with other groups, or in predator defense, for example, cannot be classified along a graded continuum. For vervet monkeys, similar referents in the external world may be designated by acoustically different calls. In a parallel manner, similar-sounding calls may designate different referents, providing further evidence against the hypothesis that non-human primate calls are simply iconic representations of the objects to which they refer. It seems reasonable to hypothesize that objects are perceived as discrete and separate entities, and that the association between a given referent and a particular acoustic unit is relatively arbitrary.

3.3. Part two

In describing playback experiments of vervet monkey alarm calls and grunts, we observers impose our own classifications on the monkeys' calls. We divide objects in the monkeys' world into discrete categories,

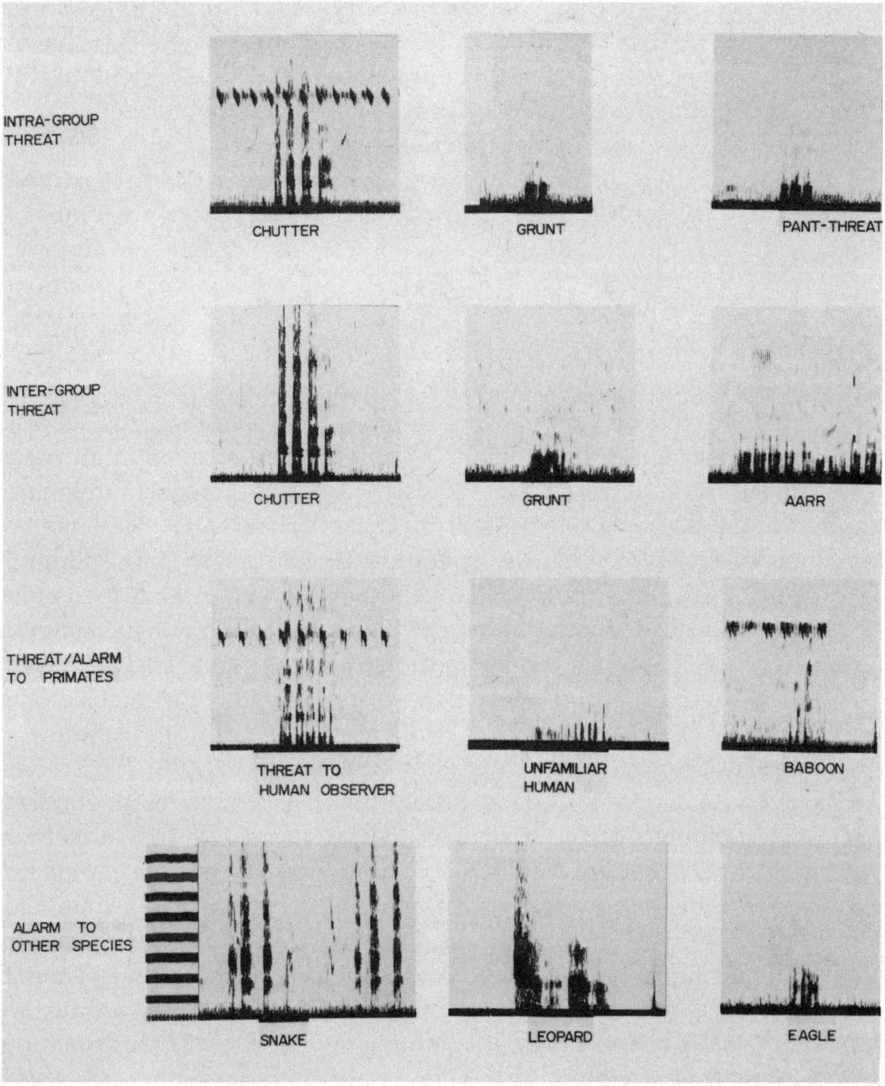

Figure 3.2. Sonograms of acoustically different vocalizations uttered in four broad contexts. Legend as in Figure 3.1.

and construct a 'vocabulary' in which we assign a given acoustic unit to a particular class of objects. Is it reasonable to suppose that the monkeys make the same kinds of classifications of objects and events in the environment? The data described above provide indirect evidence that vervets perceive objects in the external world as discrete categories, even when such objects share similar properties.

Referential signalling demands that an individual divide objects and events in the external world into discrete categories for the purpose of communication. Below, I describe in more detail what investigations of the referential abilities of non-human primates reveal about the monkeys' methods of classification. I also describe some playback experiments that were designed to examine how monkeys classify what is perhaps the most important feature of their environment – each other.

3.3.1. The ontogeny of vervet monkey alarm calls

Vervet monkeys give acoustically different alarm calls to different predators (Struhsaker 1967; Seyfarth *et al.* 1980). In so doing, the monkeys effectively classify objects in the external world. At a gross level, the monkeys distinguish predators from non-predators; at a finer level, they must determine which type of alarm call should be given to which predator. In analyzing the stimuli that elicit alarm calls, it should be emphasized that such 'category formation' may be of two types. On the one hand, categories may occur as the result of relatively simple discrimination. The monkeys may partition a continuum of stimuli into groups, in much the same way as we partition the visual spectrum into different color hues (Berlin & Kay 1969). Categories may also occur as the result of more complex hierarchical classification, such as that employed by biological taxonomists. Among humans, the creation of multilevel, hierarchical taxonomies, as distinct from simple discriminations, implies an understanding of certain formal relations between sets of objects (Kay 1971). There is some evidence that non-human primates may also be capable of understanding hierarchical relationships between sets of objects. Premack (1976), for example, was able to teach a chimpanzee not only different signs for blue, green, and red, but also the relation 'blue is a type of color', 'red is a type of color', and so on (see also Savage-Rumbaugh *et al.* 1980). To what extent do similar classificatory abilities exist among free-ranging primates, and how do these abilities develop?

Vervet monkeys in Amboseli regularly come into contact with over 150 different species of birds and mammals, only some of which evoke alarm calls from the monkeys. When infant vervet monkeys begin to give alarm calls, at about three or four months of age, they often make 'mistakes', and give alarm calls to species that pose no danger to them. What is interesting about these mistakes is that they are not entirely random. Infants give leopard alarm calls, for example, to many species other than leopards, but they give them primarily to terrestrial mammals (Seyfarth & Cheney 1980). Similarly, infants give eagle alarm calls to many species other than the martial eagle and the crowned hawk eagle, the vervets' only avian predators, but they give these almost exclusively to birds (Figure 3.3). As vervets grow older, they increasingly restrict their alarm calls to the species that are most likely to prey upon them.

The precise process whereby vervets come to restrict their alarm calls to particular predator species is not yet fully understood. Nevertheless, there is evidence that one mechanism may involve subtle reinforcement by adults. Figure 3.4 shows the number of times that infants were the first individuals in their groups to give alarm calls to different avian species, and indicates the frequency with which their alarm calls were followed by alarm calls from adults. As Figure 3.4 illustrates, adults were highly likely to follow infant alarm calls with alarm calls of their own if the infants were alarm calling at the two avian species that prey on vervet monkeys. They were much less likely to do so if infants were alarm calling at species that pose no danger to vervets.

It seems possible that adult alarm calls serve to reinforce the infants' calls, and aid the infant in sharpening its ability to respond selectively to

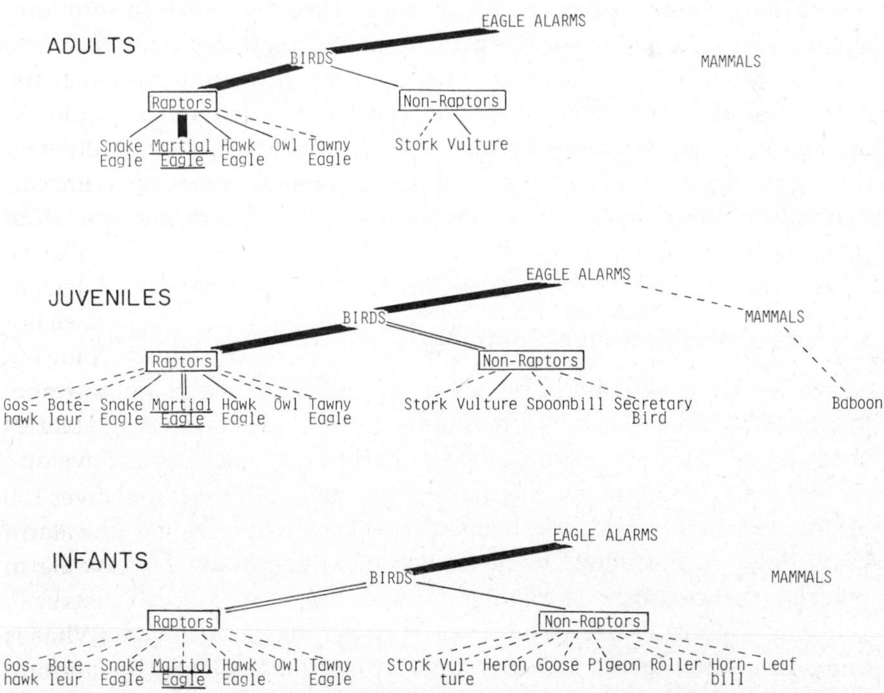

Figure 3.3. Hierarchical representation of the number of eagle alarm calls given by adults, juveniles, and infants to different avian species. Broken line, 1–5 alarms; single line, 6–10 alarms; double line, 11–15 alarms; heavy line, > 15 alarms.

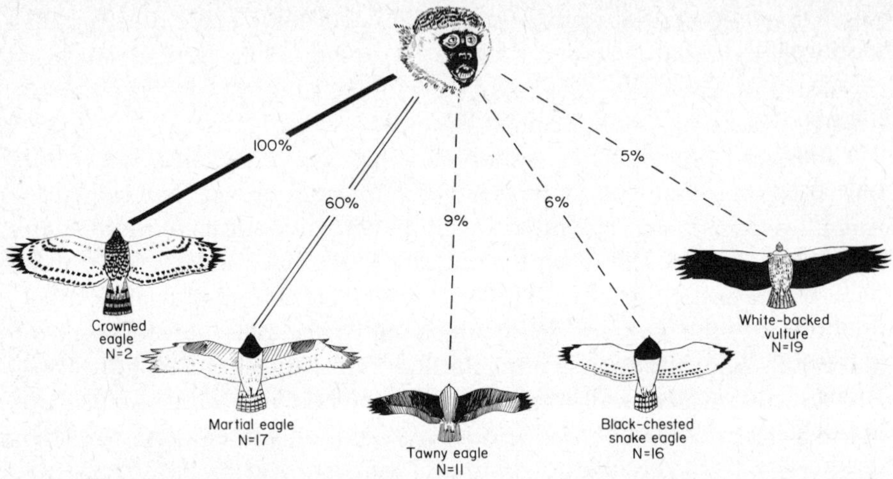

Figure 3.4. The frequency with which infant alarm calls were followed by alarm calls from adults when infants alarm called at different avian species. N, total number of first alarm calls given by infants to each raptor species.

one or two predator species within a general class. From even a very early age, vervets seem able to divide predators into broad, mutually exclusive groups, such that leopard alarm calls are restricted to terrestrial mammals, and eagle alarm calls are restricted to birds. The ability to distinguish species within broad classes, however, seems to develop during ontogeny, and there is evidence that reinforcement from adults may aid in sharpening the association between a particular species of predator and a given alarm call.

3.3.2. The classification of individuals outside the social group

The referential signalling abilities of vervet monkeys are not simply restricted to alarm calls. As mentioned above, there is evidence that vocalizations uttered during social interactions, such as grunts and chutters, may also function in a manner that effectively designates objects in the external world. In many respects, such calls are the most intriguing, for they allow the human observer to examine how monkeys perceive and classify one another.

In our attempt to understand how vervets perceive one another, we can begin at the level of the social group, and investigate how vervets regard individuals in neighboring groups. Monkey groups do not exist in social isolation from each other. Males transfer to neighboring groups at sexual maturity, and our observations indicate that the pattern of male transfer between particular groups both affects and is affected by relations between groups.

Each of the three vervet groups studied has at least five contiguous neighboring groups. In theory, therefore, males could transfer to any one of a number of neighboring groups. Instead, over the past five years, male transfer has been remarkably non-random. Males from Group A have tended to transfer to Group B, for example, and males from Group B have usually transferred to Group A. This non-random distribution of male transfer is correlated with reduced hostility between groups. Relations between groups that have exchanged males in the past are less aggressive than relations between groups that have not (Cheney 1981; Cheney & Seyfarth 1982*b*). These observational data suggest that vervets do not simply regard the members of other groups as strangers. Instead, it seems probable that the members of different neighboring groups are recognized as individuals, and that vervet monkeys recognize that certain individuals are associated with particular neighboring groups.

In order to test this hypothesis more precisely, we conducted a series of playback experiments which were designed to determine whether vervets recognize individuals in neighboring groups on the basis of voice alone. The stimulus used for these experiments is a call which is peculiar to inter-group encounters, and which is uttered only by adult females and juveniles – animals who have never lived in another group. The vocalization, called the 'long aarr' by Struhsaker (1967), and discussed above (section 3.2, Figure 3.2), is composed of a series of reiterated tonal and non-tonal units. Stimuli for the playback experiments consisted of calls from a number of different members of Groups A and C. Subjects for the experiments were members of Group B, whose range bordered that of Groups A and C. In a series of paired trials, the vocalizations of individuals from Groups A and C were played to the members of Group B, either from the range of Group A or from the range of Group C. Results indicated that subjects responded more strongly to a given vocalization when it was played from the inappropriate range than when it was played from the range of the vocalizer's own group (Figure 3.5; see also Cheney & Seyfarth 1982*b*).

These playback experiments provide confirmation of cross-group recognition by vervet monkeys. Vervets seem able to identify with which group a particular call is associated, even when they have never lived in the same group as the vocalizer. The members of other groups are apparently not just recognized as strangers, but as individuals living in particular neighboring groups. Social groups are distinguished as discrete and separate units, as illustrated in Figure 3.6.

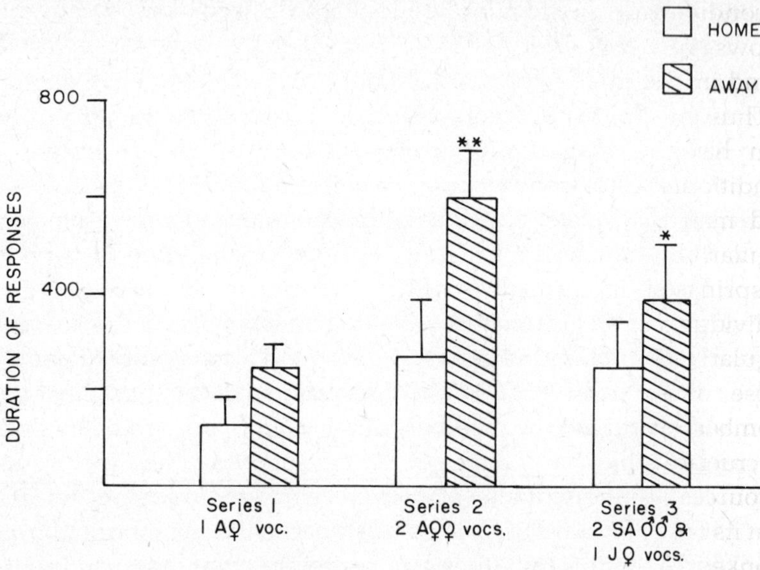

Figure 3.5. Means plus standard deviations for durations of responses to vocalizations played from appropriate ('home') and inappropriate ('away') ranges. * denotes a significant difference between 'home' and 'away' trials, using a one-tailed Wilcoxon test. Duration is measured in frames of film, at 18 fps, and indicates the length of time that subjects either looked toward the speaker, approached the speaker, or displayed in trees. The *x*-axis indicates the age/sex class of the vocalizers in each series of experiments (see also Cheney & Seyfarth 1983).

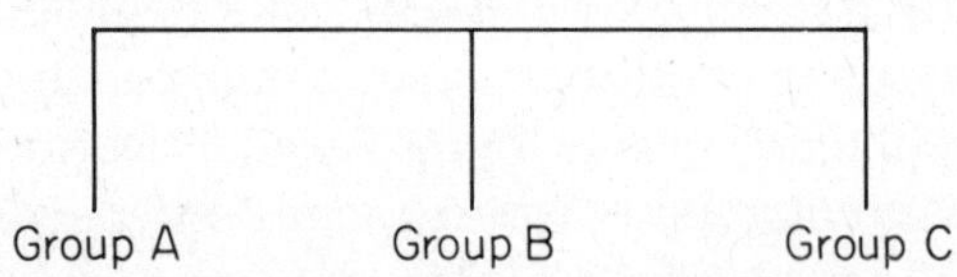

Figure 3.6. Representation of the manner in which vervets perceive neighboring groups.

3.3.3. The classification of individuals within the social group

The ability of monkeys to recognize and classify each other reflects the selective pressures acting on individuals within a particular social framework. Data on recognition therefore allow the human observer to consider social structure from the monkeys' perspective. Among human societies, social anthropologists have long used kinship terms as a tool to study how individuals classify each other, and these classifications often have two distinctive features. Firstly, they are hierarchical: group members are not just distinguished as individuals, but are arranged into higher order units on the basis of kinship, marriage, or some other factor.

Secondly, human classification is not simply egocentric: an individual knows not only how others stand in relation to himself, but also how they stand in relation to each other.

Thus far, few analogous forms of complex recognition and classification have been documented in non-human species under natural conditions. It is quite clear that monkeys, like many other species of birds and mammals, recognize the individuals with whom they associate regularly. Mothers in many species, for example, recognize their offspring. A group of monkeys, however, is comprised of many individuals with different degrees of genetic relatedness, and monkeys regularly interact with those outside their immediate kin group. Each of these other animals has different associations with other groups members, and knowledge of this complex network of relationships may be crucial to an individual's success in competing with others for scarce resources or social partners. Does a monkey recognize the relationships that its fellow group members have with each other? For example, does a monkey recognize that animals A and B are the offspring of animal C?

To investigate this issue, we conducted a series of playback experiments on maternal recognition of juveniles' screams. When conducting these experiments, we first waited until a previously designated mother was out of sight of her two-year-old offspring and in close proximity to two other 'control' females who also had offspring in the group. We then played the scream of the mother's offspring from a concealed speaker, and filmed the females' responses. When the behavior of mothers was compared to the behavior of control females, we found that mothers responded more quickly and more strongly to the screams than did controls. These responses demonstrated that female vervet monkeys were able to classify juveniles into at least two categories (offspring and non-offspring) on the basis of voice alone. Group members were distinguished as individuals, as illustrated in Figure 3.7 (see also Cheney & Seyfarth 1980).

In addition, however, there was one further finding from these experiments that was particularly intriguing. When responses of control

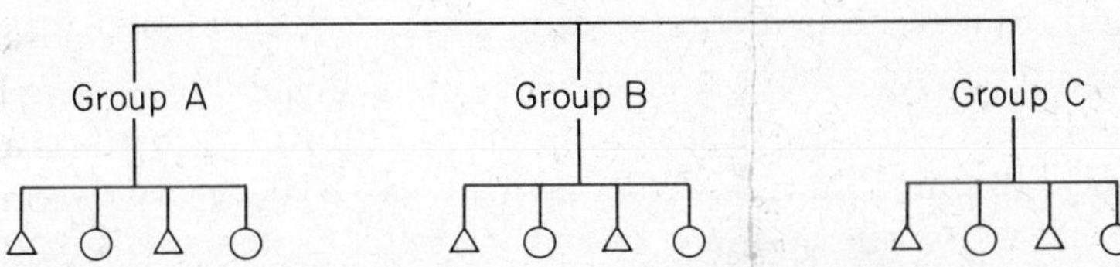

Figure 3.7. Representation of the manner in which vervets perceive individuals in their social groups.

Table 3.1. *Results of scream playback experiments*[a]

	Look after Playback	
	Yes	No
1. Did playbacks cause controls to look at mothers?[b]		
Look before playback		
Yes	4	0
No	12	16
2. Did playbacks cause controls to look at controls?[c]		
Look before playback		
Yes	7	1
No	5	19
3. After playback, were controls more likely to look at mothers than at controls?[d]		
Look at mother	12	16
Look at control	5	20

[a](1) and (2) compare the behavior of controls before and after each playback. (3) tests whether playbacks caused more control females to look at mothers than at controls. (3) eliminates females who looked at mothers or controls both before and after playback.
[b]2-tailed McNemar Test, $P < 0.010$.
[c]2-tailed McNemar Test, $P > 0.200$.
[d]$X = 3.17$, $0.10 > P > 0.05$.

females were analyzed in more detail, we found that playbacks significantly increased the probability that controls would look at the mother. There was no change, however, in the probability that controls would look at other controls (Table 3.1; Cheney & Seyfarth 1980, 1982*b*). The controls' behavior did not appear to be caused simply by cues received from the mother. Instead, control females behaved as if they were able to associate particular screams with particular juveniles, and these juveniles with their mothers. The monkeys appeared to proceed beyond simple egocentric discriminations such as 'my offspring' versus 'another juvenile' to create a hierarchical taxonomy, in which group members were both distinguished as individuals and arranged into higher order units, apparently on the basis of matrilineal kinship. This apparently hierarchical understanding of the social relationships that other group members have with one another is illustrated in Figure 3.8.

Investigations of the vocalizations of non-human primates provide numerous opportunities for the human observer to examine how monkeys classify one another. Even the subtle grunts used by vervets during social interactions reveal much about the ways in which other individuals are perceived. Acoustic analysis of the vervets' grunts has not

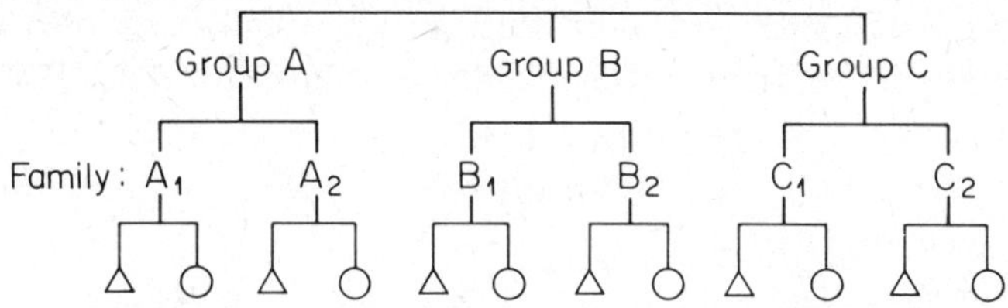

Figure 3.8. Representation of the ways vervets apparently arrange individuals in their social groups into higher-order units, on the basis of matrilineal kinship.

yet been completed, but anecdotal evidence suggests that these calls should provide much information about vervet social structure from the monkeys' point of view. One preliminary and very anecdotal example illustrates this point. As mentioned earlier, all vervet males transfer to neighboring groups at sexual maturity. There seems to be considerable variation, however, in the ease with which a male is integrated into his adopted group. While some males seem to be readily accepted by the group's females, others are actively avoided by females; they receive no grooming, and their attempts to copulate are rebuffed. Preliminary acoustic analysis of different grunt types suggests that when females grunt to males who are accepted members of the group, they give a grunt that has the acoustic properties of the grunt to a dominant (see Cheney & Seyfarth 1982*a*; Seyfarth this volume). If, however, the male has not yet been accepted by the females, the females' grunts to him will have the acoustic properties of the grunt to another group. In other words, the females' grunts reveal that although the male may be sleeping and moving with his adopted group, at least some group members still regard him as an outsider. Thus analysis of the vervets' grunts indicates that the monkeys do not perceive all group members similarly. Indeed, the grunts suggest that some individuals may be classified as members of entirely different social groups, regardless of their spatial associations.

3.4. Conclusions

Investigations of the vocal communication of non-human primates provide a means by which to examine social structure from the monkeys' perspective. Attempts to investigate the cognitive abilities of captive apes have often focused on the extent to which non-human primates classify objects and events in the external world, and act adaptively on the basis of their extrapolations (see e.g. Premack 1976; Premack & Woodruff 1978*a*, *b*; Savage-Rumbaugh, Rumbaugh & Boysen 1978; Savage-Rumbaugh *et al.* 1980). Few attempts, however, have been made to consider how the ability of non-human primates to classify objects in their environment might have evolved. It may well be that this ability derives from the selective pressures acting on individuals to make complex, hierarchical classifications of what is perhaps the most important feature of their environment – each other.

References

Berlin, B. & Kay, P. (1969). *Basic Color Terms: Their Universality and Evolution.* Berkeley: University of California Press.

Cheney, D. L. (1981). Inter-group encounters among free-ranging vervet monkeys. *Folia Primatologica,* **35**, 124–46.

Cheney, D. L. (1983) Extra-familial alliances among vervet monkeys. In *Primate Social Relationships: An Integrated Approach,* ed. R. A. Hinde. Oxford: Basil Blackwell.

Cheney, D. L. & Seyfarth, R. M. (1980). Vocal recognition in free-ranging vervet monkeys. *Animal Behaviour,* **28**, 362–7.

Cheney, D. L. & Seyfarth,R. M. (1982*a*). How vervet monkeys perceive their grunts: field playback experiments. *Animal Behaviour,* **30**, 739–51.

Cheney, D. L. & Seyfarth, R. M. (1982*b*). Recognition of individuals within and between groups of free-ranging primates. *American Zoologist,* **22**, 519–29.

Green, S. (1975). Communication by a graded vocal system in Japanese monkeys. In *Primate Behavior,* vol. 4, ed. L. A. Rosenblum, pp. 1–102. New York: Academic Press.

Kay, P. (1971). Taxonomy and semantic contrast. *Language,* **47**, 866–87.

Liberman, A. M., Cooper, F. S., Shankweiler, D. & Studdert-Kennedy, M. (1967). Perception of the speech code. *Psychological Review,* **74**, 431–61.

Premack, D. (1976). *Intelligence in Ape and Man.* Hillsdale, NJ: Lawrence Erlbaum Associates.

Premack, D. & Woodruff, G. (1978*a*). Chimpanzee problem solving: a test for comprehension. *Science,* **202**, 532–5.

Premack, D. & Woodruff, G. (1978*b*). Does the chimpanzee have a theory of mind? *Behavioral & Brain Sciences,* **4**, 515–26.

Ristau, C. & Robbins, D. (1979). A threat to man's uniqueness? Language and communication in the chimpanzee. *J. Psycholinguistic Research.,* **8**, 267–300.

Savage-Rumbaugh, E. S., Rumbaugh, D. M., & Boysen, S. (1978). Symbolic communication between two chimpanzees. *Science,* **201**, 641–44.

Savage-Rumbaugh, E. S., Rumbaugh, D. M., Smith, S. T. & Lawson, J. (1980). Reference: the linguistic essential. *Science,* **210**, 922–5.

Seyfarth, R. M. & Cheney, D. L. (1980). The ontogeny of vervet monkey alarm-calling behavior: a preliminary report. *Zeitschrift für Tierpsychologie,* **54**, 37–56.

Seyfarth, R. M., Cheney, D. L. & Marler, P. (1980). Vervet monkey alarm calls: semantic communication in a free-ranging primate. *Animal Behaviour,* **28**, 1070–94.

Struhsaker, T. T. (1967). Auditory communication in vervet monkeys. In *Social Communication Among Primates,* ed. S. A. Altmann, pp. 281–324. Chicago: University of Chicago Press.

Zoloth, S. R., Petersen, M. R., Beecher, M. D., Green, S., Marler, P., Moody, D. & Stebbins, W. (1979). Species-specific perceptual processing of vocal sounds by monkeys. *Science,* **204**, 870–2.

COMMENT

JUSTIN LEIBER

Frans de Waal has mentioned, with respect to the chimpanzees he observed, that *real* dominance relationships changed gradually while at some point toward the end of such a process of reversal one found an abrupt change in the greeting gesture that signs, or formally acknowledges, dominance. Thus A's dominance over B might be challenged with increasing success over a period of months while, nonetheless, B would still acknowledge A's dominance as usual through taking the subordinate role in the greeting formalism. Finally, after the *real* reversal was virtually complete, this reversal would be asserted formally. De Waal asked whether this happened with vervet monkeys.

As I construe her, Dr Cheney has responded by saying that something like the reverse occurred. Among vervets the subordinate gives a *grunt-to-a-dominant* when greeting a dominant while the dominant is likely to be silent. Dr Cheney remarked that the first observed change between a dominant X and subordinate Y might be simply that, when Y gave the grunt-to-a-dominant to X, X also gave this grunt. She said that it was likely then, within a few days, that the *real* dominance relationship would clearly begin to change. She went on to remark that she presumed that the real dominance relationship had begun to change some time before the simultaneous-grunt occurrence but in such a subtle way that the human observers did not notice. When I questioned this, Dr Cheney agreed that perhaps one ought not to make this presumption and that it would be most interesting to determine whether the simultaneous-grunt, rather than an acknowledgement, or description, of an already-proceeding change in real dominance relation, might itself be the first, or an important, step in this change.

We have been warned by linguists such as J. R. Firth or philosophers such as J. L. Austin that speaking is as much an action in the world as a picture, representation, or comment about the world. But we need such a warning because we are inclined to forget that speaking is action and we may need this warning too in investigating non-human primate signing. De Waal's talk of the *real* dominance relationship as opposed to the formal acknowledgement in the greeting gesture might make the uninitiated forget that a change in the greeting formalism might be as real and as powerful a move as a physical blow. Dr Cheney's initial presumption, that there must have been some prior though humanly unnoticed behavioral change before the simultaneous-grunt, ignores two interesting possibilities. One is that there was no prior behavioral change, nothing observable in vervet behavior by either human or vervet, while indeed there was an internal psychological or physiological change in the formerly dominant vervet, an internal change that led to the unexpected grunt-to-a-dominant which occurrence itself might cause further reversal behavior. The other is that there was no prior behavioral or internal change whatsoever, the unexpected grunt-to-a-dominant being an accidental byproduct of unrelated matters like a genuine slip of the tongue. Though the behavior might in this manner be an accident, it might nonetheless be a contributing cause (even in some way an initiating cause) in reversal. To put the matter in human terms, one is often conscious that one has made a slip, or said something unintentionally bright, witty, stupid, or malign; but the response of others follows so closely and irredeemably that one becomes responsible for what one has said.

It would seem that this phenomenon is of the utmost significance for research into communication and signing among primates. This is so for at least two reasons. The first, that those most skeptical about supposed primate signing are inclined to see such behavior as an automatic reflexive response to other internal or behavioral activity. If this signing behavior, on some occasions, must be explained as planned or strategic behavior, or accidental behavior, then this skepticism is shown to be wanting. Second, it is just this sort of phenomenon that shows us that signing communication is a central dimension of the primate's existence. If accidental or planned signing can significantly alter the activity and status of a primate, then the ability to identify, produce, and strategically manipulate signs would be understood as an identifiable and highly selective biological feature.

Should the proper technology of micro-loudspeakers become available, it would be most interesting to determine whether one could alter the life course of a primate by providing it with a more powerful and aggressive 'voice' or the reverse.

Part II
Theoretical preliminaries

Introduction

Two main problems are addressed in the theoretical aspects of this study. What implications are there for our view of animal thought from the widespread use of intentional language to describe the signalling systems of non-human primates? Closely related to this discussion is the issue of the influence of ways of speaking developed for specific human purposes and in specific human communities for talking about primate communication.

Leiber, Harré and Harris share the view that intentional ways of speaking cannot be eliminated from primate studies, but they also agree that this fact, in itself, has no implications for any hypotheses about the quality of animal mentation. Leiber's argument turns on the point that cognitive psychology is based on hypotheses about the existence of 'processing modules'. While the ape-language projects demonstrate a high order of semantic capacity (calling for intensional attributes to primates) they go no way to proving a syntactic capacity, without which it cannot be said that apes 'have language'. So whatever cognitive activities depend on linguistic capacity ought not to be ascribed to primates.

By a rather different line of argument Harré shows that ethologists are constrained to use concepts appropriate to goal-directed action. In consequence, the form of any cognitive psychology of primate interaction must involve the supposition that apes use means–end thinking. It may even be proper to use intensional hypotheses, that is to attribute conceptual thought to primates. But that does not imply conscious deliberation. By distinguishing between the public-collective structural and functional units of a communication system and the alleged private-individual mentation of the creatures using it, Harris is able to define the limits of constructive anthropomorphism, as restricted to the inter-individual communication system in public use.

Ardener's very close look at the vocabularies developed for describing primate signals brings out the absolute dependence of even apparently onomatopoeic expressions on their specific linguistic origins. Anthropomorphism is not just a matter of the semantic implications of the lexicon, but also is inherent even in the phonetic conventions under which the sounds of animal communication are mimicked. Asquith distinguishes between specific anthropomorphic theses, those which ascribe identical cognitive processes or affective states to primates as to humans, and general anthropomorphism. The latter amounts roughly to the idea that animal actions (including signalling) should be explained in a general framework of psychological theories and concepts that have the same *general* character as those we use for understanding human actions. Only by using intentional language can functionally identical behaviour units be made to fall into the proper categories of action rather than mere behavioural descriptions. Unlike Harris, but like Harré and Leiber, Asquith argues a case for treating anthropomorphic terms as metaphors, on the basis of which 'second order' action, roughly cognitive activity (for instance goal-setting), can be ascribed to those primates for whose actions the vocabulary of intentional description seems the only appropriate means of classification.

While there is a large measure of agreement among the authors in this section about the necessity for the use of intentional language in the description of primate signalling systems, there is some diversity of opinion as to how far animal cognition should be treated in terms of 'intension', that is as the use of concepts. All are agreed, however, that the necessity for intentional modes of description has no implications whatever as to the phenomenology of animal mentation.

A. Intention and action

4. The strange creature

JUSTIN LEIBER

Vernon Reynolds has suggested that the contributors to this volume fall into two groups: those interested in primatologists and those interested in primates, that is, theoreticians and empirical researchers. Here I would like to stress my non-membership of either group. I am not an empirical researcher studying either human or nonhuman primates. I have, however, been an animal trainer, working with mountain lions and, after a fashion, with chimpanzees. I have also had this relationship with human primates. So you may take it, invoking the anthropologist's vocabulary, that I have been a participant with a variety of animals though not an observer. As a Professor of Philosophy I might claim the role of theoretician, but I do not want to take up that mantle here. Rather, I will claim in addition to the role of animal trainer that of novelist. In this case the novel is more about the dramatic clash of ideas than of individuals, or perhaps of ideological persons as opposed to particular persons, but I shall talk about some primate individuals and in any case my theme will be the disappearance of the whole thinking organism into component organisms. Thus, this is a French novel.

It is a commonplace of this volume and of recent work on animal cognition in general to distinguish an intentional and mentalist mode of describing animal psychology from an automatic, reflexive, or behaviorist one. As many have pointed out, the same distinction has been made regarding human cognition. Some, though perhaps no contributors to this volume, would maintain that the automatic behaviorist account is the only scientific mode for describing either animals or humans. Some, though perhaps few of the present contributors see the intentional, mentalist account as appropriate only to humans, denying that this mode is ever acceptable or scientific or true as regards nonhuman animals. And many, and perhaps many contributors to this volume, see the intentional, mentalist account as appropriate only to organisms that have language. Some, finally, have looked at the primate research, particularly the ape language research, of the past decades, as a theoretical and empirical morass, complaining that the notions of mind and language, of meaning, sentence, and communication, are hopelessly unclear, so that the theoreticians should just stop talking about labels and theories and methodology and let the stalwart researchers get about their job. Alternatively they believe that, in the spirit of the recent *Clever Hans Conference* (Sebeok 1980), what is needed is an unmasking of the supposed researchers as charlatans who deceive themselves and others, as seance-givers and conjurers. Proceeding in a catholic manner, I wish to deny all of these claims and to impugn above all the dichotomy with which I started.

In particular, I believe that the ape language research of recent years has been an extraordinarily profitable interaction of linguists, those studying artificial intelligence, philosophers, anthropologists, and primatologists, an exemplary interplay of theory and praxis. This is, *also*, a German novel.

It is, to put my principal thesis, also a German novel in a dialectical sense. For we will find sufficient neither the view that all nonhuman and human primate behavior is automatic-causal, nor even the view that some is so but some is intentional-mental. Rather, we need to distinguish the linguistic from our other bio-neurological component organs and from holistic talk of mind, intention, communication, meaning *and* automatic. We have come to see that we need to separate three things: (*a*) the automatic, stimulus-response, or quasi-causal (*b*) the mental-intentional and (*c*) the linguistic. While I am here dramatizing, or novelizing, the substance of what I maintain is now plainly the view of two of the most distinguished empirical researchers in this area, Professors Herb Terrace and David Premack: viz., that chimpanzees in particular do not have the essential features of the human language organ but, nonetheless, have a rich and complicated mental-intentional life. Of course, may I add, it has also long been the view of Noam Chomsky, who has argued since 1955 that our linguistic capacities, our autonomous, formal, phonological-syntactical organ, must be understood as quite distinct from much of meaning and reference, from our other conceptual, motor, and perceptual activities which we may largely share with other animals, and from organism-wide talk of thought, planning, emotion, and personhood which, again, animals may merit as much as we do. To put the matter in rough empirical terms, the chimpanzee may be understood as the Broca Syndrome human, like the language-deprived wolfchild *Genie* whom Susan Curtiss has characterized for us (1978).

Stretching matters further, we might see the Wernicke Syndrome human as a phonological-syntactical computing machine that lacks the primate conceptual-perceptual-motor apparatus and that uses the phonological-syntactical machine as a way of exponentially expanding the thinking power of the primate, or primitive, mind.

Let me recapitulate this extraordinary stretch of discovery by reviewing the work of Beatrice and Allan Gardner and the work in linguistics that has played off against this, particularly the work of Ursula Bellugi and Ed Klima on American Sign Language. Since I have adopted the novel form, the narrative-historical form, we have a sequence of episodes in which we are gradually forced to distinguish the stimulus-response, the cognitive-intentional, and the specifically linguistic.

The narrative form is particularly appropriate here in that it is a boot-strapping operation all the way, theorizing suggesting experimentation, experimentation suggesting theoretical reconsideration, and that in turn new experimentation and critical reinterpretation of old results, and so on.

The question 'What is the essence of language?' had no clear answer at midcentury. If, in part as a result of the work we review here, we have more of an answer today, we can hardly maintain the matter is settled. What has always seemed clear is that humans at least have language and that it is central to human cognitive and social function. Hence it was wholly natural for the Hayeses, in the 1940s, to attempt to determine chimpanzee language capacity simply by bringing up a chimpanzee as an ordinary human child. Indeed the Hayeses might be said to have followed what philosophers recently have called a 'realist' view of basic terms in empirical science (Kripke 1972; Putnam 1975). The realist takes it that terms for natural kinds like 'water', 'heat', and in this case 'language', designate underlying essential properties, though humans who use these terms may often know their application through superficial phenomenal correlates (Leiber 1975). Thus, so the realist account goes, the AD 1500 speaker who said 'water' and 'heat' in fact meant H_2O and the mean motion of molecular particles, though this speaker could not give our rendition of what he meant; this realist account has the happy result that we can maintain that scientists *discovered* what water and heat are, not that they changed the meaning of some terms (*mutatis mutandis* for the ancient Roman, and so on).

We take the same position for 'language'. The Hayeses could surely not say what is essential to language, perhaps not even whether its essence is to be sought in some functional, social, structural, or biological feature. Nonetheless, they plausibly proceeded, for they knew where language was located – in the normal maturation of the human in a normal human maturational environment – and they simply sought to see how the chimpanzee infant would fare in similar circumstances. For this procedure to work, one must presume that chimpanzee and human infants share much in motor, perceptual, and social capacities but this has seemed a reasonable presumption. More recent experimenters – the Gardners, H. Terrace, and F. Patterson – have followed this realist bent. Indeed, as I shall later suggest more explicitly, the realist account is the one that makes best sense of the bootstrapping between theory and experiment.

How did the Hayeses' infant chimpanzee fare? Notoriously, the chimpanzee developed many perceptual, motor, and problem-solving abilities much as the human child, and often earlier. But at best the chimpanzee could vocalize less than a half dozen English words after several years of effort (Hayes & Hayes 1951). This now familiar result was taken by many at the time to suggest a vast gap in general intelligence between chimpanzee and human. As I shall argue it may rather suggest a modular psychology within which one can say that the chimpanzee has much of our mix of cognitive components but lacks something homologous to our language organ. R. A. & B. T. Gardner (1975) suggested a third explanation. They equated a rich intentional-mental life with having essential linguistic capacities, in effect denying modular psychology or at least a strong version of the autonomy of syntax. They came to suspect that chimpanzees had linguistic capacities but lacked a voice box. Chimpanzees were mute. The Gardners attempted to prove their point by teaching Washoe what they took to be American Sign Language (ASL). Their point clearly would be established if (*a*) ASL is essentially equivalent to other natural languages and if (*b*) they succeeded in teaching chimpanzees ASL. Obviously something could be made of partial success. Much of their point would be established if, at least, (*c*) chimpanzees learned much of the essential features of ASL and learned them in much the way of a human child. I will argue that the work of the last twenty years on human signers has established point (*a*) but this very work, coupled with the work with chimpanzees, has made (*b*) and (*c*) quite difficult to maintain. The research story is, of course, quite a complicated one in that much of what we now know about ASL was not available until the seminal work *The Signs of Language* (Bellugi & Klima 1980). Further, for unfortunate practical reasons, it has not proved possible to employ fluent ASL signers as primary, long-haul teachers and companions for the chimpanzees. The chimpanzees have learned from pidgin signers as Professor Terrace has remarked (personal communication).

Washoe and subsequent chimpanzees trainees in 'ASL', after years of daily training, have made the rough acquisition of a sign vocabulary of one or two hundred words. The vocabulary is similar in meaning to that of two- to four-year-old humans: chimpanzees share many interests in food and fun with human children. The chimpanzee uses this vocabulary to get various kinds of food and drink, to initiate chasing, tickling, and grooming. What goes on can be described as one- and two-gesture communication acts (Peterson 1980). Longer strings seem not to be sentences but rather jumbled repetitions. Nonetheless this research has suggested that the Gardner's signing chimpanzees have quite outstripped the Hayeses' oralizing chimpanzee.

This contrast is an illusion.

Let me begin to show this through a very simple example, one which initially is only likely to impress those who have not signed with chimpanzees. The chimpanzee is reported to sign MORE DRINK. To sign MORE, one brings one's two fists together a foot or so in front of one's chest. To sign DRINK one makes a right-hand fist, thumb tip protruding a little beyond the index finger knuckle, and then one touches one's lip with the thumb tip, almost as if one might be mimicking the act of drinking (one's fist representing a cup). In this crude, pidgin version of ASL one can teach a human to sign MORE DRINK in a few seconds. Think, for comparison, of the length of time one might take in teaching a human, wholly unfamiliar with the phonology of English and its vocabulary, to *say* 'more drink', or, for that matter, to *hear* these words as distinct from 'more ink', 'more think', 'nor blink', and so on.

Further, in anything like mature human communication MORE DRINK is elliptical for any of the following: '*Pour me some more of that drink, I want another cup of that, There is some more drink over there, Do you want more to drink?, Is there some more to drink?, Shall we have more to drink?, Give us some more to drink*'. Indeed in pidgin ASL we would also have: '*There is a lot of drinking going on, He is drinking lots, There are cups over here*', and so on. In pidgin ASL, DRINK can be both noun and verb, cloaking a range of subject, object, and inflexional process, and MORE can cover from 'very' to 'several'. One is obviously tempted to exaggerate how much gets said and to attribute much to 'contextual meaning'. How much of this is in any sense a matter of *language*? Certainly, the DRINK gesture seems iconic. And one could argue that if the chimpanzee gestures with its empty cup with orange juice in the offing, this is equally significant and equally non-linguistic as pidgin MORE DRINK in the same circumstances. What is one to make of this? Is the Hayeses' chimpanzee, who was (as most of a highly social species) quite given to complicated social interactions of a non-linguistic sort with human interlocutors, that far behind?

Indeed one might begin to wonder about human signing, about the status of ASL itself. There are several points that have been made in arguing that ASL and sign languages generally are not comparable to the more familiar natural languages. It has been claimed that sign language by contrast with natural language is (*a*) largely iconic, (*b*) non-abstract and heavily context dependent, (*c*) devoid of syntactical resources and form words, consisting of concrete signs that contextually function as nouns or verbs, (*d*) universal and natural, easily picked up by common sense as an extension of ordinary gestural activity. With respect to fluent ASL signers these claims clearly appear to be false (Bellugi & Klima 1980). These claims are not wholly misleading however.

ASL is not now iconic in any very significant way. The language was recorded, taught, regularized, and in part invented during the French Revolution, coming to the United States in the second decade of the nineteenth century. When one looks back at early versions of ASL, though in part this must be a product of conscious pedagogy, invention, and fanciful etymology, the iconic element seems present in many basic signs. But many of the traces of this have disappeared over the years from ASL. In present ASL, the remaining supposed iconic elements seem wholly opaque to the uninitiated (this also speaks to the notion that sign language is universal and transparent). Further, short-term memory tests make it clear that fluent signers hardly make any use of supposed iconic elements, nor do they much depend on meaning. Rather, signs are stored as packages of distinctive formal features, a sign formally consisting of one of a small number of hand configurations, one of a small number of hand movements, and one of a small number of points of articulation on the signer's body and the space in front of it. Memory mistakes occur because of formal (quasi-phonological similarity), and not for semantic or iconic reasons.

Indeed, the fluent ASL signer has a rich set of formal quasi-inflectional devices that are laid on top of the basic signs. For example, Bellugi & Klima (1980: 264–6) show that the basic SICK sign can be inflected by formal flourishes indicating predispositional, 'susceptative', continuative, incessant, 'frequentive', intensive, approximative, and resultant aspect, not to mention other devices that make clear whether the signer, signee, or some other is the subject of the ascription. *Cogito ergo sum* is more the spirit of ASL than 'I think therefore I am'.

If a fluent signer signs MORE DRINK it can be perfectly clear which of the many possibilities mentioned above is meant quite apart from context. Bellugi & Klima convincingly argue that ASL has rich formal productive processes that give it the abstraction, detail, and expressive power of any natural language, though it lacks the huge vocabulary of English and word order plays little role in its syntax. Looking at the apparent history of the development of ASL, Bellugi & Klima suggest that what one sees is the progressive disappearance of iconicity, natural gesture, and context dependence, and the growth of syntactic and quasi-phonological processes, almost as though the autonomous language organ were re-creating the formalisms of speech in another medium. This would echo a favorite view of Eric Lenneberg's about the development of spoken languages and the re-creation of their grammars in successive generations of speakers (Lenneberg 1967). Bellugi & Klima also note that much of what they have demonstrated about the syntactical process of ASL is normally no more consciously available to its fluent speakers than the syntax and phonology of a spoken language is to its speakers. The notable prejudice against sign language has delayed its serious scientific characterization until quite recently.

This may also account for two overly ambitious claims about sign language that have appeared recently. One is the claim that children exposed to signing learn their first sign words some months earlier than more ordinary children learn their first spoken words (Williams 1976). Reflecting on the point that one may teach an uninitiated human to pidgin sign MORE DRINK in a few seconds while considerable time may seem to be needed for a similar vocalization, one might plausibly conclude that *initial* signing, which is clearly gestural and context-dependent if not iconic, is not distinctively *linguistic* or at least not comparable to the formal linguistic achievement, the command of distinctive vocalizing features, required for the spoken 'more drink'. More formalizing needs to be done, so to speak, to produce an acceptable vocalization, though in a short time this apparent discrepancy disappears, and we find the linguistic development in the human child of ASL and English running in strict parallel. This interpretation is precisely what one would expect if we were seeing parallel homological development of a discrete organ system molding different input/output media to the same inner requirements.

Much the same may be said of the other claim. In a liberal reading of *Beyond Herodotus* one could take the authors to be claiming that children born deaf who are not exposed to signing, will spontaneously invent sign language among themselves, developing this invention at the same pace as hearing children exposed to a spoken language (Gleitman *et al.* 1976). The gestural communication that these four- and five-year-old children developed was iconic, context-dependent, transparent to the uninitiated, and lacking most of the formalisms of ASL, though doubtless an isolated community of such signers would over the years, or generations, have developed a fully-fledged language comparable to ASL.

What is clear, to return to the chimpanzees, is that they show, after years of training and exposure to signing, not the slightest trace of homological development parallel to that of human children. With a few months' training they grasp the communicative use of basic signs, or at least their use in terms of getting food etc. This might be thought similar to the gestural prelinguistic behavior of one to two year olds I mentioned above. But the chimpanzees, as Professor Terrace does so much to show in this volume, show no tendency to go beyond this, no tendency to pick up the formalisms or syntactic processes of ASL, no tendency to expand and play with the formalities of language, no tendency to be engaged by the properties of language for its own sake. Indeed word order, though emphasized perhaps through misunderstanding, is by no means the most important of these. If we think of words as bearers of phonological and syntactical processes, that is, discrete levels of distinctive features and inflections after their counterparts in a real sign language, then it is simply not the case that the chimpanzees have learned words. So far as language acquisition goes, regarded as quite separable from general intelligence and social know how, the Hayeses' chimpanzee continues to be the norm.

I will conclude with three points.

Note how the realist account of scientific terms makes our bootstrapping account of the interaction of theory and experiment so appropriate. The Hayeses, and latterly the Gardners and Herb Terrace, properly began by assuming that they knew where language was located – in the normal maturation of the human child in normal maturational circumstances – and sought to see how the chimpanzee would fare in such circumstances. The limited performance of the Hayeses' chimpanzee suggested an investigation of chimpanzee success with ASL, which in turn led (along with other factors such as Chomsky's claims regarding the autonomy of phonology and syntax) to intense research into fluent human use of ASL. This was paralleled by Terrace's concerted effort to look for formal features in chimpanzee signing. This energetic interaction of theory and experiment leads us to a secure sense that we are *discovering* the essence of language, not arbitrarily inventing it or legislating the meaning of the English word 'language'. A realist, or demonstrative, account of scientific terms is what allows us to make proper sense of the bootstrapping of theory and experiment.

One notes also how important it is to make the distinction between homology and analogy, between genetic/developmental/species-wide accounts and functional ones. No doubt the chimpanzees communicate. They achieve some of the basic functions of language. But the underlying homology is not there.

And this leads to a further moral point. The account in which it seems that either chimpanzees function at an automatic level or have a rich intentional/linguistic life is no longer forced upon us. We can maintain that chimpanzees have such a rich cognitive/social/emotional existence without feeling forced to say that they have language. To recall my claim that the language organ (by itself a sort of formalizing moron), exponentially increases the capacity of the primate conceptual/perceptual/motor system; we may conclude that what sets us apart from our fellow apes gives no guarantees of goodness, happiness, moral status, or human solidarity. But then, looking at our twentieth century, I take it we may perhaps find this a sober but not startling conclusion.

Let me add a word about the notion of intentionality or intentional characterizations of animal activity as this has surfaced here. In the sense in which it has most plausibly been used in this volume, to characterize talk about animals as intentional is to say that the talk in question correctly describes the animals in question as having intentions, as having a mental life such that we have to describe it as consisting of propositional attitudes such as *believes, knows, suspects, imagines, questions, hopes, wishes, fears, etc., that thus and so* (one fills in the *thus and so* with any of a range of propositions). Logicians, however much they have deplored this necessity, are agreed that propositional attitudes and most terms in propositions require intensions in addition to extensions. *This,* for example, may be understood as one of those fairly rare extensional terms in that it just stands for something pointed to in the local environment: the term has no meaning (*Sinn* in Frege's vocabulary), no intensional or conceptual content.

On the other hand, most terms, all the way from *cat, water, baby, tree, running, threatening,* etc. to *believing, knowing, science, necessary,* etc. are intensional. To say, for example, that the chimpanzee Washoe 'believes that there is a cat behind the sofa' let alone believes that this is a picture of a cat, is to describe Washoe's mental life as including intensions. I have no doubt that Washoe's mental life has such intensions, and that hence it is intentional. What is worthwhile in the Gardners' work cannot be understood in any other way. However, this does not commit me to the view that Washoe has language or that it is at all plausible to think that Washoe has, as we do, on some occasions an inner theatre of introspective self examination and self consciousness. As Rom Harré has pointed out, the base level ascriptions of an intentional life to animals (and often to humans) are person- or organism-wide, depending in part on species and social community for their sense, and in no useful way requiring the inner theatre. I believe that a modular psychology is the only one in which we can make real empirical sense of the notion of inner theatres, as, that is, the interfaces between modules such as languaging and imaging. And I do not believe that language is the only medium in which such self consciousness is available, though language surely looms large in our mental life. But these last points cannot be argued further here.

Someone might ask what all this is supposed to amount to and what suggestions it contains as to how research should proceed.

There really is something to Rom Harré's point about not changing languages, or discourse levels, within the same experimental model. There are legitimate ways of describing us that are automatic and simply causal (this is also true of other animals and of computers and robots). And there are equally legitimate ways of describing us that are intentional and simply mental (this is also true of other animals and of computers and robots). This second level of intentional description does *not* at all require that we attribute *free will* or superphysical causal powers to ourselves (or to other animals, or to computers and robots); nor does, as I have remarked, this intentional level of discourse require postulation of an inner theatre of self conscious self description, though this inner theatre does come in with addition of a language organ.

Indeed, someone may still ask why it is so necessary to keep levels distinct, rather than using any system that works, reporting everything that looks to be true and not worrying about segregating it into different bundles. But that, it has been the burden of my remarks, is where the Gardners went wrong. It is also where Bellugi & Klima went right. The Gardners did not see that they could describe Washoe's intentional mental life without saying that this meant Washoe was saying sentences of sign language to herself. They were insensitive to the formal apparatus of signing as distinct from its communicative and referential use, insensitive to the quasi-phonological and syntactical processes natural to human, but not chimpanzee, developmental programs. Bellugi & Klima, on the other hand, found again and again that even fluent speakers of ASL were often unaware of the rich formal means that they were employing in the everyday business of communication and practical activity. In their experiments they had to isolate formal linguistic process from the surrounding emotional, perceptual, contextual, and cognitive processes with which it functioned.

When I say of a man (*a*) 'he is angry that she upstaged him' I am not committed in any clear or important way to saying he is saying (or has said) to himself (*b*) 'I am angry with her for upstaging me'. Indeed to confuse these is a confusion of discourse levels for (*a*) describes a propositional attitude, a general state of mind, and (*b*) reports a time-datable event, an interface process between different modules, and has an autonomous and formal structural character. Description (*b*) reports the mental occurrence of a *sentence of English* and requires the existence of the English language to make sense. Description (*a*) reports no occurrence of any linguistic sort whatsoever, nor does it require the existence of any particular human language, or indeed human language generally, to make sense. In case (*a*) I may say that that is his propositional attitude – that that description makes rough good general sense of his mental life – and it might at the same time be the case that he has never actually said to himself that he has this attitude. In this last case we imagine that he will be *startled* to hear the sentence 'I am (you are) angry with her because she upstaged me (you).' But he cannot be startled by his propositional attitude alone! He can only be startled by the explicit linguistic event of an occurrence of the description of his propositional attitude. If the man is angry that she upstaged him, he cannot (logically cannot) not be angry that she upstaged him. But of course he *can* be startled to be told, or think to himself, 'You are (I am) angry that she upstaged you (me).' Now the point about non-linguistic but cognitive-intentional individuals like chimpanzees is that they cannot, or mostly and most significantly cannot, be said ever to be in a condition of this sort.

To return to the theme of this novel, I invite you to continue this most exciting interaction of theory and experiment.

References

Bellugi, U. & Klima, E. (1980). *The Signs of Language.* Cambridge, Mass. Harvard University Press.

Curtiss, S. (1977). *Genie.* New York: Academic Press.

Gardner, R. A. & Gardner, B. T. (1975). Early signs of language in child and chimpanzee. *Science*, **187**, 752–3.

Gleitman, L. & Goldin-Meadows, S. (1976). *Beyond Herodotus.* Philadelphia: University of Pennsylvania Press.

Hayes, K. J. & Hayes, C. (1951). The intellectual development of a home-raised chimpanzee. *Proceedings of the American Philosophical Society*, **95**, 105–9.

Kripke, S., Davidson, D. & Harman G. (1972). 'Naming and necessity.' In *Semantics of Natural Language*, ed. Hittikka *et al.* Dordrecht: D. Reidel.

Leiber, J. (1975). *Noam Chomsky: A Philosophic Overview.* New York: St Martins Press.

Lenneberg, E. (1967). *The Biological Foundations of Language.* New York: Wiley.

Peterson, P. (1980). *The Apes and Language* Houston Conference 1980.

Putnam, H. (1975). The meaning of 'meaning'. In *Minnesota Studies in the Philosophy of Science,* ed. K. Gunderson, Vol. 7, pp. 131–93. Minneapolis: University of Minnesota Press.
Sebeok, T. A. & Rosenthal, R. (eds.) (1980). Clever Hans Conference. *Annals of the New York Academy of Sciences,* April 1980.
Williams, J. S. (1976). Bilingual experiences of a deaf child. *Sign Language Studies,* **10**, 37–41.

COMMENT

D. PLOOG

Leiber's remarks on the sign language of the deaf (ASL) clearly show the vast difference between the signing of apes and the signing of congenitally deaf people. I learned from Ursula Bellugi and Edward Klima that sign language is a fully developed language with all the linguistic properties of a spoken language, prosodic features of language included. In neurological terms this means that the 'language organ' of which Leiber spoke is detached from the vocal-articulatory system in the brain since it is operating in the visuo-motor mode and not in the audio-vocal mode. It is nevertheless based on biological foundations: the primary acquisition of ASL takes as long as the development of spoken language. There seem to be 'mile-stones' of language as in spoken language. There also seems to be a critical period: children who learn ASL after, say, seven years of age, will have a sort of foreign accent phenomenon, as Eric Lenneberg called it; they will not speak like native signers. That signers also can become aphasic has been known for a while but is still poorly understood.

COMMENT

DUANE QUIATT

I am not convinced that applying the terms 'analogous' and 'homologous' to mental properties is likely to improve understanding of similarities and differences between ourselves and other primates. I suspect that use of these terms may entail unnecessarily roundabout comparisons and subsidiary terms that are more abstract than need be. In biology, 'analogy' and 'homology' lost most of what value they once may have had as *explanatory* concepts with the elaboration of a genetic theory of evolution. Biologists, once they have got beyond the introductory comparisons of whales, fish, bats, and men, may not agree as to where it is useful or even appropriate to distinguish between similarities that are analogous and those that are homologous, or what is to be gained (in clarity of description) from making such distinctions.

Comparisons in terms of analogous function and homologous structure are most problematic in the case of close phylogenetic relatives. In such cases, like structure may or may not indicate like function, but we cannot expect to find the kind of superficial resemblances between functionally important structures that would call for explanation in terms of convergent evolution. Similarly, while identical structure may afford a wide range of differences in species-characteristic behavior, similarities in important features of behavior between close phylogenetic relatives are unlikely to derive from different structures. The problem is that the framework of comparison, if we apply terms like 'analogous' and 'homologous' to species recently divergent, is not dichotomous, as it must be for such terms to be useful even in a descriptive way.

5. Vocabularies and theories

ROM HARRÉ

5.1. Vocabularies and their effects

It is widely agreed that some form of linguistic determination of the patterns of thought must be admitted – that features of the language one uses determine in some measure what one can express with it and so, to a degree, limit the uses one can make of it. Theses of linguistic determination run from the extremes of the Sapir–Whorf hypothesis – that the actual forms of thought and perception are fully determined by the available linguistic resources – to more modest claims about the pre-determination of limits to thought, of vocabulary-linked preferences among possible perceptual identifications selecting from all that we could, in principle, experience, and so on. I do not wish to defend the strong claim.

There is some variability too in what might be covered by the term 'linguistic'. To restrict it to the writing and speaking of words is almost certainly too narrow, since many other things can be used for purposes very similar to those for which we use words (gestures, diagrams, pictures, and so on). I would rather talk of semiotic than linguistic determinism – though the examples I will be discussing are, in the nature of the case, linguistic, since I am proposing to examine a vocabulary.

What could choice of a particular vocabulary pre-determine? There are two basic theses involved in conceiving of such a pre-determination, of whatever strength it be. There is the thesis of 'theory-loading' – that there are, or could be, no terms in a language which have purely descriptive content. In the weakest form, 'theory-loading' suggests that a term, merely by being used within a grammatical category, presupposes a general metaphysical theory – in the strongest that some terms are so theory-dependent for meaning that their use as descriptions presupposes a particular scientific theory. For instance, the semantic and syntactic rules governing the use of the word 'horse' require a shared metaphysical theory of a world of persisting material things shared between speaker and hearer. To talk of lightning as an electrical discharge is not only to describe the phenomenon in different terms from those of commonsense, but to introduce into that description the principles by which it would be explained, via the meanings of the words 'electrical' and 'discharge'. So, choice of vocabulary not only affects what we pick up from our experience, but may also determine the means by which it could be explained.

Correlative to the thesis of theory-loading is a thesis about the effect of theory-change. Introducing a new theory does not leave the meaning of the descriptive terms used with the old theory unchanged. So a change of theory for an empirical domain must be accompanied, at least potentially and partly dependently on just what kind of theory-change occurs, by a change in the empirical selection of 'facts' from the domain as we experience it. Some recent writers have developed this thesis into an extreme form of relativism. They have argued that facts must always somehow be in accord with theory, at least with the current general theory appropriate to an empirical domain, since it is that theory which partly determines the meaning of the terms which are used to identify and thereby to select those matters of fact to which we can pay attention. Within a theory-defined domain, facts to refute the theory could not be identified. Of course, small scale theories, within a larger theoretical framework, are falsifiable by contrary facts. But large scale theories can be displaced only by rival theories, which permit a new structuring of experience, coordinate with the new loading of the descriptive vocabulary.

Most commentators would now agree that in the extreme form, this thesis has been over-generalized. One could perhaps partition the rules of use, even for heavily theory-laden terms, say, in our context, a term like 'adaptive strategy', into those related to the general metaphysics of the domain, closely related to that of the commonsense categorization of the world, and those deriving from theories of various degrees of specificity. The term 'strategy' would be related to a more general theory, made more specific with the qualification 'adaptive'. Thus, adherence to a primarily evolutionary and genetic theory as the main explanatory device, would lead to the adoption of terms like 'adaptive strategy' to describe activities within the chosen domain. And the use of such a term would yet further presuppose that there were individual animals interacting somehow, which is a generic, metaphysical theory. The latter theory would be a common part of the semantic field of both the neologism 'adaptive strategy' and the archaism 'cunning move' which it replaces, since both are applicable in the same metaphysical framework, individual animals interacting, and *both* would be inappropriate to, say, a group-mind metaphysics of primitive troops.

5.2. The role of scientistic rhetoric

The semantics of descriptive vocabularies is made much more complicated by virtue of the use of specific vocabulary, not for what one might call pure scientific purposes, but as rhetoric, as part of the techniques of self-presentation. Lay persons reading ethology and psychology, for instance, often mistake this kind of useful display for pretentiousness, but it is nothing of the kind. C. Bachmann & H. Kummer (1980) sometimes slip in rhetorical uses of terms that would otherwise be improper. For example, on page 317 they say, 'A male . . . has tactile access to the female'. What they meant was 'A male . . . could touch a female'. Why use 'had tactile access'? By shifting from the verb 'to touch' to an abstract noun 'access', the *metaphysical level* of a discourse is raised from the concrete to the abstract, and by using the term 'tactile' instead of the ordinary 'touch', the *technical level* of the discourse is raised from lay to scientific. The rhetorical effect is to present Bachmann & Kummer as not just anybody 'looking' at baboons but as workers in a technical-abstract enterprise, scientists 'observing' baboons. And this is a useful reminder of their own status *vis-à-vis* their engagement with these entertaining animals. It is also worth noticing that they use this scientific vocabulary almost exclusively when they describe their empirical work; and this is just the context where the issue, layman or scientist, can be raised. By using a certain rhetoric they present themselves as (make a claim to be) on the 'scientific' side of the dichotomy.

5.3. Theoretical frameworks

The development of ethological vocabularies not only illustrates the use of a rhetoric of scientisms to create an impression of professionalism, but the conditions for that very impression depend upon an interplay between two radically different theoretical frameworks (at about a middle level of generality). I shall call these the 'causal framework' and the 'intentionality framework'. Choice between them is loosely related to the currently acceptable level of anthropomorphism deemed proper in animal studies; this is a point I shall return to (see also Asquith, this volume). At first (say twenty years ago) the scientific vocabulary of ethology drew heavily on the vocabulary of the intentionality framework (e.g. the writings of Lorenz 1966 and Thorpe 1966). More recently (say fifteen years ago) a mistaken positivism led to an attempt to create a descriptive vocabulary with no theoretical implications other than simple Humean causality (e.g. Blurton-Jones 1967). Of course, such a task is logically impossible, and all that could be achieved was a descriptive vocabulary loaded heavily with the assumption that animals were automata – one way of grounding a Humean analysis of the production of action as stimulus and response. (Human psychology is still struggling with the similar, and even more perverse error, though in the human case adherence to the mistaken philosophical theory of operationalism was compounded by a political motivation.)

More recently still it has been realized, not only that the dependence of ethology on a philosophical theory as crass as empiricism (operationalism) is a mistake, but that conceptual resources of animal studies can be greatly enlarged by judicious use of intentionalist hypotheses. I want to trace the growth of this idea in following the changing theoretical loadings of the vocabulary of ethology over recent years. Intentionalist hypotheses are rather general, being compatible with many different forms of specific 'mentalist' hypotheses, of which cognitivism is only one. In this chapter I am concerned to trace only the most generic forms of intentionalist theory – leaving the science of ethology to determine the level and form of mentalism needed for explanatory purposes, though I shall make some suggestion in section 5.6. on intentional and intensional modes of description.

My argument is best represented in a three-dimensional Cartesian space. One axis represents the contrast between explanations couched in intentional language and those couched in Humean causal language. The second axis represents the distinction between explanations which involve hypotheses of explicit cognitive processes such as remembering, deciding, etc. and those which presuppose only that animals are complex automata, for the description of which cognitive terms are metaphorical. The third dimension marks the contrast between hypotheses requiring the existence of conscious cognitive processes, like those of human beings, from those which are agnostic, call them a-conscious.

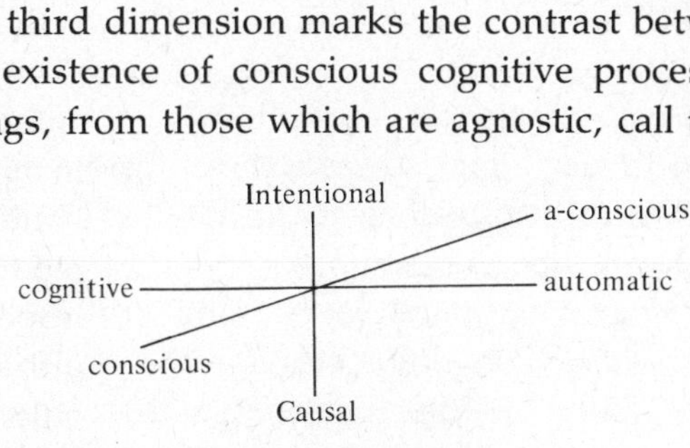

Figure 5.1. Kinds of explanation.

I argue that a study of the uses of language by ethologists suggests that their intuitions as to the nature of that which they are observing in studying the lives of primates force them towards intentional-cognitive hypotheses. I argue further that the use of the language of intentions and cognitions does not commit an ethologist to any particular view about the states of mind of animals, for instance whether they are conscious or not. I happen to think that many animals are conscious in much the same way as we are, but that none are self-conscious. To set out the arguments for that belief would take us too far afield and it is not germane to the points I want to make in this chapter.

5.4. Acts and actions

I propose to discuss the issue of acts and actions in terms of the distinction between three ways of partitioning and classifying what primates (and that includes people) do. Attempts have been made to partition their streams of activity into units of behaviour, mere movement, but there are no stable taxonomies of movement that have proved worthwhile. If classified as actions, the activity-stream is partitioned in accordance with the completed behaviour modules that seem to then form an animal doing *something*. The most refined partitioning derives from some idea as to what (act) the animal is doing (performing) – and this requires the observer to treat the animals as forming a society. The most sophisticated partition would begin 'at the top' with social categories (acts), using those to identify doings (actions), which are themselves performed by some disjunctive class of physical movements (behaviour). It is vitally important, I believe, to grasp the last condition, that the same act-action can be behaviourally realized in this *or* that *or* the other movement. I have seen no evidence that suggests that acts-actions as behaviours are in 1 : 1 correspondence, even among sticklebacks.

Suppose we jump forward now to look at the consequences of having already made the transition to act-action description vocabularies. What is implied about the proper mode that the explanation of action should take? Insofar as there were pure behavioural descriptions, allegedly, they were loaded with biological theory – and so were directing their users towards explanations by reference to causal mechanisms. It is not worth discussing any of that in detail. But the action-vocabulary necessarily involves implicit reference to an intentional background to the action. Just how that background is to be converted into specific explanatory theories is a matter for detail – but on pain of incoherence they cannot be causal theories. To be able to act intentionally is to be able to think. So correlative to an action-vocabulary for describing and partitioning actions, there must go a cognitive (and perhaps emotive) theory of how the associated thinking is carried on. This need not be an individualistic theory, assigning the whole structure of the deliberative cognitions to individual action. It could well be that much of the thinking is collective. (The Gardners reported a chimpanzee hunt where there seemed to be social activities that one could most readily explain as collective cognitive activity.)

5.5. Ethologist's vocabularies as a 'transitional' terminology

These considerations make the following passage from Bachmann & Kummer (1980:315) extremely puzzling. They say 'The causal interpretation is that baboons may be capable of assessing aspects of relationship among two other members of their group'. But 'assessing' is just the kind of activity one would call *cognitive* rather than *causal*. The simplest hypothesis, that they mean 'explanatory' by 'causal', cannot be sustained since they also say 'potential contestants should not only weigh the asymmetry in dominance (which is one aspect of resource holding potential, RHP), but also an asymmetry in "pay-off" which the prize will yield to either of them'. 'RHP' must be a causal concept – while 'weigh' is an intentional-cognitive one. Perhaps it is the intentional vocabulary that is metaphorical here, and really the authors are still subscribers to the old paradigm. I hope to illustrate in more detail the *transitional* nature of their conceptual scheme in a detailed examination of their vocabulary. The quotations above, if every term is taken literally, show that they are using items which are loaded with an intentional-cognitive theory, together with others which are loaded with its rival – a causal automaton theory. The difficulties in sustaining this combination as *all* literal uses, can be seen by referring to Figure 5.1.

This conflict of usage is all the more puzzling in that Bachmann & Kummer (1980:316) adopt an uncompromisingly cognitive approach. They say 'the male's *information* on pay-off asymmetry concerns other variables than the preference of the female. These variables are *unknown* to the experimenter, but *known* to the males. . .' (my italics). Experimenters know 'variables' cognitively. So if 'know' preserves its meaning in the first and second halves of the sentence, the males must be doing relative to these 'variables' what the experimenters might have been doing, but were not, namely 'knowing them'. In either case it seems to be 'information' in the ordinary sense that is being talked about, and the structure of argument entails that it is the human conceptual system that is at work here.

To bring out the 'transitional' character of the semantics of Bachmann &

Kummer (1980) I analyze one further passage. Writing of the relation between experimentally induced and natural behaviour they say (p. 318) 'animals in the three "roles" display the same behaviours as in the original field and enclosure experiments where the baboons were less restrained. This should alleviate doubts as to whether the behaviour of our baboons was "natural", that is, where they "recognized" the situation as that of a rival facing a pair and responded with an evolved and adaptive mechanism'. Note first the use of scientistic rhetoric, e.g. 'displayed the same behaviours' for 'did the same things'. The signs of the transitional character of this passage are twofold. Scare quotes are put around 'roles' and 'recognized', presumably to show that these are being used non-literally, but still they *are* being used. However, to say that the baboons responded with 'an evolved and adaptive mechanism' is to adopt an uncompromisingly biological explanatory framework, as opposed to a cognitive or intentional one. But the authors do not sustain their biological explanatory framework as we shall see when I analyse their own further account in detail.

In the key paragraph of their article, Bachmann & Kummer (1980:320) use a variety of terms giving their preferred account of the results of their studies. These terms fall into two main categories, one of which can be further sub-divided.

The first category of terms are wholly non-cognitive, suggesting the operation of causal, and automatic mechanisms. This category is made up of the terms 'inhibit', 'release' and 'elicit'. Further, it is not the creatures themselves that 'elicit', 'inhibit' or 'release' their tendencies, but attributes such as 'unknown characters that make them attractive, etc. etc.'. There can be no question but that these are causal terms. It is worth noticing that the theory of causality implicit in these usages is non-Humean, that it is not a mere empirical regularity that is cited as causal, but the inhibition or release of a tendency. So even though this aspect of the explanation of baboon behaviour is automatic and causal, there has been an advance on the simplistic Humean causality of purely behaviourist explanations.

The second category of terms is explicitly cognitive and since examples of it are used without any scare quotes, I suppose we are justified in taking the terms literally and in presuming that the authors do mean to ascribe to baboons the kind of cognitive processes they seem to ascribe. This category falls into two sub-categories – terms used for deliberation and decision, and terms used for cognitive-affective states relevant to deliberation. So in the first sub-category we have 'assess', 'prefer' and 'decide', the verbs for modes of thought, and in the second, 'unwilling', finding something 'attractive', 'choice' (that is, what has been chosen). In the use of these three terms we have a sketch of a process, not of the inhibition or release of tendencies, but of modes of thought. These are fairly unsophisticated modes of thought, not calling for complex or long-sustained cognitive processes, but to 'assess', 'prefer' and 'decide' are methods of reasoning with which we are ourselves familiar.

Now the question is what to make of the presence of these two categories of terms, apparently implying radically different kinds of explanation. Is a resolution possible under which they can co-exist? Or, have our authors simply found themselves poised in a transition from one conceptual scheme to another, and not quite ready to abandon the old one?

The evidence strongly suggests the second alternative. It seems to me clear that there is no way for reconciling causal with intentional explanations, *within the same conceptual system.* In a causal process, the elementary phenomena are related by causal mechanisms, so laws of nature describing the regular behaviour of such mechanisms are necessary only relative to the stability and persistence of the mechanisms, and not by reference to the nature of the phenomena they relate. So, from a description of a cause, one cannot, in the absence of any other information, deduce a description of its usual effect. This is why Hume's famous theory of causality (as mere repetition of like pairs of events) gets a purchase. If one is sceptical about our right to knowledge of unobserved causal mechanisms, all that is left for a scientist to work on are sequences of logically independent events and conditions. But when phenomena are semiotic, they are related in a meaning system and the case is very different. The sense of an action (for example, its social force) is partly given by the intentions of the actor, partly by the interpretations of the interactor. So what act-action is performed in making a movement is partly defined by actors' intentions. A general statement describing regularities between intentions and actions involves a conceptual relation, just because the act-action, as the meaning or significance of a behaviour, is partly determined by the actors' intentions, and thereby partly defined by it. In consequence, a description of a movement as the act-action which follows a certain intention, provided it is not in some way aborted or inhibited, could be deduced from a description of that intention. If I intend to insult you, then my gesture is, from my point of view, necessarily an insult. Further, it follows that in an intentional system there is good reason for ascribing an intention to an actor that is semiotically related to the action he/she/it performs. There could never be a perfect match, since part of the meaning of an action is the interpretation made of it by the interactor. You may not realize that my gesture was insulting, but in general semiotic interaction is possible as a sequence of coordinated actions from each contributor if intention and interpretation have a reasonably good match.

If the tendency to the use of intentional language became established, two consequences would immediately ensue:

(*a*) The level of generality usually required of a scientific account would be achieved only with the semiotic description, and the 'laws' would express semantic relations rather than regularities in the behaviour of causal mechanisms.

(*b*) The relation of the semantic system to its biological base would be contingent and come under the principle that philosophers call 'token-identity'. This means that while each semantic element is realized in some physiological/ecological element, the next repetition of a similar semantic element, in the next interaction, or in another animal, or both, could be realized in a physiological/ecological element of a different kind from that in which it was realized before. Types do not map onto types – only tokens onto tokens.

These principles, if true of baboon society, would put a limit on how far baboon social behaviour could be explained genetically. To put this another way: to the extent that these principles are true of baboon social activities, baboon societies are cultural phenomena. Insofar as they are cultural phenomena, competence in social action would require education over and above maturation – and would open up the possibility of cultural differentiation among groups of conspecifics, as well as the possibility of innovation. Arguing the other way, if there is cultural differentiation and if there have been observable innovations, then it is reasonable to conclude that the creatures have evolved a semiotic level of interaction, that is they have a system of public discourse in which behaviour as actions *stand for* acts in such a way that the 'stand for' relation is conventional.

Ambiguity reflects the claim that creatures act intentionally and have a mental life. The double-sidedness of this claim infects even the case of human beings. If the tendency I claim to have detected in ethological work – as represented in Bachmann & Kummer (1980) – is indeed real, then the human case is providing the essential source-model for the development of a scientific account of primate interaction. But the source-model is double-sided. On the one hand one means something private and personal by talking of the mental life; on the other one means something public and social, essentially engaging in discourse with one's fellows. It is my view (following Wittgenstein 1953) that the latter is prior and that the former is a late development or appropriation (cf. Vygotsky 1962) of the latter. So in suggesting that the evidence implies that animals have a kind of mental life, I am not proposing that they have a kind of secret life – an inner theatre in which simulacra of the world are reproduced; they may, they probably have, but that is irrelevant to the issue of the nature of social interaction. The point is, rather, that they display their intentions in their actions, and generally are capable of operating with conventions, that is to say with a double system – the signifier and the signified. And *that* is, I think, what emerges from the gentle slide in the works of ethologists so beautifully exemplified in Bachmann & Kummer (1980). This reflects the perception that baboons, like human beings, live in a world that is partly constituted by signs and symbols.

5.6. Intentional and intensional modes of description

My argument so far has established a strong tendency amongst ethologists to use intentional language in describing the phenomena they observe. We can decide whether this tendency should be resisted or indulged only if we have a clear idea of the various levels of commitment that a user of intentional language engages. Furthermore, ethologists can be reassured that there is nothing unscientific about the use of intentional descriptions, provided that their commitments are appropriate to the subject matter of the description. To be scientific in one's descriptions is, amongst other things, to be in command of a vocabulary with which one can make precise distinctions (and numerical measurability is not a necessary condition for precision), and to be able to specify the conditions both for using and for refusing to use a particular term. Many of the spontaneous uses of intentional language by ethologists meet these conditions.

But intentional language is part of a larger conceptual structure of modes of description, and by specifying this larger structure one can come to see how uses of intentional language can be related to levels of commitment to theoretical or unobservable entities, such as individual mental representations. For the purpose of this chapter I distinguish only three levels in the hierarchy of intentional languages, general goal-directed action talk (GDA), intentional language proper (IL), and cognitive descriptions. All three have in common the following features:

(*a*) Activities are identified by dual criteria, the one as to their immediate characteristics, and the other as to their upshot. All forms of intentional description then are of a means-end form. Scientific uses of GDA at whatever level presuppose *some* form of prior representation of the goal state; but general GDA does not, so far as I can see, in any way entail that the prerepresentation of a goal state is mental, and certainly it does not entail that it is conscious.

(*b*) Descriptions of activities in intentional language either explicitly or implicitly involve reference to the end or goal in the predicates descriptive of the action. This is the linguistic or semantic aspect or consequence of the criterial feature described in (*a*) above.

With the general category of GDA descriptions I propose to specify a narrower category of intentional descriptions by reference to the relation that obtains between the end and the means for its realization. If ends and means are related only by physical causality I suggest we should refuse the intentional description. We should use it only if we have reason to believe that ends and means are related by convention. Our reasons for holding that, in a particular case, the relation of means to ends is conventional would be, for instance, late learning and cultural diversity. A special case would be where the means is a sign for the end, as it may be that a gesture will direct a creature to something which does not currently exist, or is not present. To decide whether the 'bee-dance' should be described in IL, we would need to decide whether the relation of dance-pattern to the action required to find the honey is in some degree conventional.

The most general form of intentional description would be act-action descriptions, where the act was defined in terms of a social or some other upshot, the action was the means used to perform it, and where action and act are actually related by some non-mechanistic link such as convention or knowledge. An act of warning could be performed either by an upraised hand (paw), or by a grimace, or a chattering, etc. Thus predator warnings as uttered by birds would fall under a general GDA description, but not be properly described in strict intentional language, while some at least of the repertoire of chimpanzee actions ought properly to be described intentionally. But intentional language involves a commitment, namely to the consequential ascription of intentions (defined in act terms) to the creature performing the action. A more tightly specified form of intentional action description now emerges. That is where the mediating link between action and act is to be thought of in terms of knowledge. There are many cases of general GDA description where a mediating mechanism, rather than knowledge, is presupposed in the use of means-end descriptions. When we say 'The chimp did this because it knew that by doing this it would get the banana (frighten its rival), etc.', we are committed to the hypothesis that chimpanzee competence is properly to be described as knowledge.

In this way, a gradual tightening of the conditions for using intentional language draws us towards ever stronger commitments to a cognitive account of the genesis of action. At this level of specification we are committed to the idea that a propositional, or at least a conceptual, account of animal ways of acting is implied. It is crucial to realize that it may be quite proper to use cognitive (conceptual or propositional) modes of description of the preamble to action when it would be improper to ascribe states of consciousness to the creatures so described. At least the cognitive mode implies no more than that there are modes of organization of the neural mechanisms that control action that have structures not derivative directly from biophysical processes, but superimposed on them, in much the same way that programmes are superimposed upon the electronics in machines.

Where would such modes of organization come from? The main source of such structures in man is the public-collective discourse mediated by language and other symbolic systems. These have interpersonal and emergent properties, which, if the likes of Vygotsky (1962), Mead (1934) and Wittgenstein (1953) are right, determine the overall cognitive structures of individual thinkers, by superimposing a non-physiological metastructure upon the neural basis of action. One would expect to observe the same thing in primates to which one wished to ascribe intentions in the strong cognitive mode, but to whom, for one reason or another, one was unwilling to ascribe consciousness.

To talk of conceptual modes of organization as opposed to and imposed upon physiological modes, introduces another concept, 'intension', often confused with the notion of intention with which I have discussed the issues of action description so far. The intension of a class is the concept (or in linguistic cases, the description) under which any member must fall, while its extension is the set of members which satisfy that requirement. Intensions are concepts, extensions are sets of things. According to the view I have been elaborating above, at a certain level of specificity of description actions are both intentional (that is, goal-directed by virtue of a cognitive structure pre-existing the action and representing the outcome of it), and intensional (that is, directed by a structure whose components are those higher order entities we call concepts, and whose relations are sematic or conventional, and in my view ultimately social). At any rate, they are not simply elaborated forms of physiological relations.

My suggestion of specifying more and more stringently the conditions for using GDA forms of descriptions yields a hierarchy of explanatory concepts rather similar to Dennett's (1978) recent suggestion of a set of levels of intentionality. In both cases the hierarchy provides a succession of increasingly stringent ways of specifying what is to be thought to be interposed between the conditions of action and the performance of action. As I have argued the matter it turns out to be a double hierarchy, involving both the form taken by prerepresentations of goal states to be achieved, and the form or status of the relation between means and end so represented. The double hierarchy is needed since there may be cases where, although the goal representation is cognitive (conceptual), the means are related to the realization of the goal by non-conceptual mechanisms. There may be other cases (say hunger) where the goal is set physiologically, but the means for its achievement are, at least amongst human beings, related through a conceptual system, which may include such matters as religious taboos, social status and the like. So I envisage a hierarchy which has some zig-zags and dog-legs in its design, but for which the most stringent level requires intensional, that is conceptual, relations to be involved in both goal setting and goal realization.

If my general line of analysis is right, it involves a doubling-up of the states of actors and interactors. And one would expect there to be a correlation between neurological complexity and degree of stringency in the proper application of intentional/intensional language. And that is just what we find. The most complex neurological thing we know is a human being, and it is to humans that intentional/intensional descriptions, with all their implications, apply with the least controversy. The phenomena which lead students of the lives of other primates also to use intentional/intensional language seem also to match up levels of intentional/intensional description of the behaviour of beings relative to the complexity of their neurological systems.

5.7. Intentions and representations of goals

In adopting a general intentional 'framework' for describing primate activities, one is determining the 'level' at which a search for pattern will be conducted. In particular, by the use of intentional modes of description, one will be directing attention to the possibility of seeing practical activities as networks of actions towards realized and *unrealized* goals. It is the possibility of the latter that relates intentional descriptions to some kind of cognitivism, since it is natural to assume that the mode of being of unrealized goals is conceptual. So intentional descriptions lead to intensional attributions. But the psychological implication, that *each* animal has a goal or intention, ought to be drawn with care. There are many cases of human action where it seems right to attribute the goal-representation of intention to a collective rather than to each of its individual members. There may still be intentional action, though no one member has an adequate representation of the goal towards which the actions of the members are severally directed. It may be that in a traditional society there are strict role assignments, and that beliefs about the point of actions in role are widely differentiated amongst different role holders. Yet an anthropologist may be able to assign functions to the activities of each role holder that subserve some collective goal. He, the anthropologist, correctly uses GDA language but it would be impossible for the role holders to do so within the framework of their given beliefs.

However, with that caveat entered, the use of the intentional framework does allow one to formulate hypotheses about the intentions of individual animals, in the sense that they may properly be said (in circumstances to be delineated) to have prior representations of a state of affairs to be brought about. Intention is, as it were, the future tense of that of which memory is the past tense. The animal that did not disclose that it knew how to arrange the boxes to get a banana, in de Waal's film, and who acted immediately the coast was clear, represents the kind of case where not only intentional description but cognitive attribution seems right. However, the use of an intentional framework does not tell its user what form a prior representation will take. Even in the case of de Waal's chimp, though we seem obliged to use cognitive concepts to describe the event, such as 'knew', 'planned', etc., this usage could not be taken to entail that that chimp was conscious of what it was intending to do in the way that we might be. Though *I* would be quite ready to say that it was, the use of intentional language and even of cognitive concepts does not *require* such a hypothesis about the mode of prior representations.

5.8. Summary and conclusions

An intentional framework has another aspect – it admits the possibility of meaning, since what an animal meant to do is the goal of the activity for the animal. What another animal takes the activity to mean is, in some cases, identical with what the first animal meant by it. In human affairs, the meaning of an action, the social act performed in doing the action, is usually conventionally rather than naturally correlated with it. This is one of the respects in which human social life is an artefact.The use of the intentional framework allows the issue of whether any given sign is conventionally or naturally related to its meaning to be raised. If naturally related, the sign is part of a causal nexus (and the mechanisms at work in that nexus remain to be discovered); if conventionally related, the sign is part of a semantic nexus, properly part of an intentional framework, and it is the sources and moments of the establishment of conventions, rather than the discovery of causal mechanisms, that would be the next step in a research project. If meaning arises by convention, then it is something that has to be learned, and so is culture relative in principle. So it is likely to be both culturally differentiated and subject to modification and innovation in stable biological conditions.

Intentional frameworks can never be reduced to causal frameworks, *a priori*, because the structure of a pattern of action discerned by the use of this framework may be realized by different causal processes on different occasions, with the same animals or with different animals. If this is as generally true of primates as it is known to be true of humans, then the *only* general scientific framework for discerning and describing social patterns will be the intentional one. There will be a disjunctive set of causal mechanisms associated with a single intentional description. So similar activities directed towards similar goals may be realized in this *or* that *or* the other causal mechanism. Unities are semantic not causal.

Methodological Postscript 1

An analytical model or cluster of analytical models permits one to discern pattern in an empirical field of a science. An explanatory model is usually required to enable one to formulate hypotheses about causal mechanisms. Explanatory models are usually based on well articulated source models. Models in the sense I have been using the term in the above are concrete analogues. To use a dramaturgical metaphor in the analysis of social events, as Goffman does, is to use an analytical model. In supposing that a social event is like the staging of a drama, certain features of that event stand out and others are suppressed. But it would be a misunderstanding of the role of analytical models, and a confusion with the role of a model as an explanatory device, to think that the stage analogy entailed that ordinary persons *produced* their actions in the same sort of way as actors produce theirs. We must always treat the degree of coordination between analytical model and explanatory model as something to be explored and considered.

It is a desideratum for theorizing in the mature sciences that analytical and source models from which explanatory models derive, should be coordinate, that is should be using the same sorts of concepts. I can illustrate this from Darwin's *Origin of Species.* In reading the *Origin,* one is struck by the way Darwin seems to be looking at the entire organic world as if it were, to some unspecified degree, like a stretch of rural England. The pigeon-fanciers, stock-breeders and gardeners see the populations of animals and plants under their care in terms of lines of descent, and of adaptation to the demands of farm, garden and race course. All of these lines are continuously being modified by selective breeding. Now think of all nature as an English county, and one has the theory of natural selection. Part of its great power to convince us lies in the coordination of analytical model – 'lines of descent' – used to discern patterns of interest in the animal and plant worlds, and its source model – 'domestic selection' – from which the explanatory model – 'natural selection' – derives. It is clear from the first two chapters of the *Origin* that the analytical and source models are part of the same overall 'picture' of nature, with the help of which patterns are being picked out and explained.

In the case of ethology, the attraction of coordination will have an effect on the meaning of the descriptive vocabulary, through the kind of explanatory model in use, and its source model – roughly whether these are causal or intentional – stimulus and response or GDA.

Methodological Postscript 2

In social psychology there has been a persistent confusion between two notions of group or collective, and we should beware of this confusion creeping into ethology. In one sense, a group consists of all those creatures which have similar properties to one another. We often use the metaphor 'sharing' for this: 'A nation is a group of people who share a common heritage.' When European social psychologists wanted to break away from the assumptions of American individualism, they urged a shift of emphasis to the study of groups (Tajfel & Israel, 1972). But in Tajfel's (1978) subsequent intergroup studies, only groups in the above sense figured, that is his groups were constituted by him (rather than nature, so to speak) by virtue of similarity relations.

There is another sense of group or collective more germane to the interests of social psychologists and ethologists. That is that of a group as a structured and bounded entity. Such a group may have system characteristics, that is involve internal processes by which its integrity is maintained. In such groups individuals usually play distinctive and mutually supporting parts. In a structured group, one would expect each individual to be different from every other, for instance in skills and knowledge, having those appropriate for his/her role, in a specific location in the structure. In a taxonomic group as described above, every individual is necessarily in the relevant respects exactly the same as every other; quite the opposite is the case with structured groups. (The pilot, navigator, engineer, chief steward, etc. must know different things from each other.) The importance of structured groups follows from the possibility that there may be emergent properties of group action, and that the system constituted by the differentially acting unit-individuals may be able to perform activities (including cognitive processes) of which each individual, taken separately, is incapable.

Moving to the intentional/intensional modes of description should be seen as not only enriching the scope of hypotheses about individuals, but also opening up research into system behaviour of that group of individuals as a true collective, as a structured group. Sometimes it may be only the group that is the proper subject of predication of intentional/intensional concepts.

References

Bachmann, C. & Kummer, H. (1980). Male assessment of female choice in hamadryas baboons. *Behavioural Ecology and Sociobiology*, **6**, 315–21.

Blurton-Jones, N. (1967). An ethological study of some aspects of social behaviour of children in nursery schools. In *Primate Ethology*, ed. D. Morris. London: Weidenfeld & Nicholson.

Dennett, D. C. (1978). *Brainstorms*. Hassocks: Harvester Press.

Lorenz, K. (1966). *On Aggression*. New York: Harcourt, Brace & World.

Mead, G. H. (1934). *Mind, Self and Society*. Chicago: Chicago University Press.

Tajfel, H. (ed.) (1978). *Differentiation between Social Groups*. London: Academic Press.

Tajfel, H. & Israel, J. (1972). *The Context of Social Psychology*, London: Academic Press.

Thorpe, W. H. (1966). Ethology and consciousness. In *Brain and Conscious Experience*, ed. J. C. Eccles. New York: Springer-Verlag.

Vygotsky, L. S. (1962). *Thought and Language*. Cambridge, Mass: MIT Press.

Wittgenstein, L. (1953). *Philosophical investigations*. Oxford: Blackwell.

COMMENT

H. KUMMER

I very much enjoyed reading Harré's stimulating review, and I accept the criticism of mixed language which we used in our article on baboons. Nevertheless, I find it difficult to accept the advice to commit myself entirely to intentional language, for the reasons described below.

The language mixture in our paper expresses our ignorance of whether the baboons in our experiment are more correctly said to 'respond to' or to 'assess' the female's behavior toward her mate. The experiment was not designed to answer this question. Mixed language was an admittedly awkward but honest solution. In the ethological tradition, words are labels, not explanations. I have the impression that Harré attributes to words the dignity of hypotheses, an attitude which I do not share. Surely, the language we use implies a great deal, but not all language implies what biologists would accept as a 'theory'. Where, in the following series of substitutes, is the transition from one language to another: 'The animal walks and stops near A; approaches A; approaches A rather than B; prefers A to B'? I would prefer to approach the distinction on the level of explicit hypotheses and experiments rather than (prematurely, I think) on the level of language. Until experiments have been designed which determine whether our baboons 'respond' or 'assess', I see no justification in committing myself to one or the other language, particularly because the two descriptions are not mutually exclusive.

Take the fictitious example of a female bird behaving as if she knew' how well her mate will feed her young, for example by laying fewer or more eggs. She might 'know' this in very different ways, for example, by a nutritional trace of the amount of food the male fed her during courtship, or from observing how effectively he forages. We are not very interested to decide in advance whether her state should be called knowledge. Rather we wish to find out by experiment exactly what and how she knows, and until then, use 'to know' as a preliminary label in an everyday sense. Of course, even everyday words must have some 'theoretical' implications. The 'theory-load' of the word 'is' must be tremendous; should we therefore not use it in a scientific description?

My plea here is that everyday language must remain available for what has yet escaped a satisfactory explanation. If a discipline begins to use an everyday word in a scientific sense, it cannot ask others to invent a new expression to replace its everyday usage. Rather, it should create its own scientific term.

Harré's and Dennett's hierarchies of intentional languages comprise important guiding concepts for ethological research. However, the dialogue will have to continue before many ethologists can properly handle such concepts in their research.

COMMENT: TWO EYES OF SCIENCE

FRANS B. M. DE WAAL

Being in complete agreement with the suggestion that 'conceptual resources of animal studies can be greatly enlarged by judicious use of intentionalist hypotheses', it was disappointing to learn that Harré proposes to apply these hypotheses, or rather the language associated with them, indiscriminately to all animal behaviour.

It is true that primatologists, consciously or unconsciously, hesitate between mechanistic and cognitive explanations. The one is traditional and generally accepted; the other is intuitively attractive but problematic, since as yet it lacks a systematic methodology. Development of an adequate terminology and a set of methodological rules may allow us, to a very limited extent and only in theory, to enter the animal's inner world of intentions, expectations, thoughts and feelings. To deny or accept the possible existence of such a world is indeed a philosophical choice. On the other hand, whether we will make the step to a post-Cartesian ethology is not fully disconnected from empirical evidence.

We might view the animal mind as a postulated entity, comparable with our model of evolution or the atomic model (Griffin 1976:59). Although both atoms and evolution escape direct observation, we adhere to these models as long as our data fit the hypotheses derived from them. Similarly, we will never know how much our subjective experiences resemble those of animals, or even whether they have any, but we may postulate some thought processes and emotional states (and interactions between them) and see whether this brings us any further than the stimulus–response model of the black-box era (which is becoming increasingly unsatisfactory). However, if the new theoretical framework would fail to improve the elegance of our interpretations and the predictability of animal behaviour, its chances of scientific success would be low.

Science looks at animals with both a philosophical and an empirical eye. Harré is right about the growing sensitivity of the first eye for horizons which have been forbidden for some time, but he seems to underestimate the decisive influence of the other eye. For him the choice is a purely philosophical matter, hence the suggestion to apply intentional language to *all* animal behaviour. Would it really be 'judicious' to adopt a language which would not discriminate between the way Köhler's chimpanzees reached bananas and the way fruitflies do?

It is difficult to make sense of Köhler's tool-use experiments (Köhler 1917) or of my own observations of coalition-formation among chimpanzees (de Waal 1982) while remaining within the so-called 'automaton framework'. Maybe it is not impossible, but the explanations will be of an *ad hoc* character and be very 'heavy', i.e. involve a number of apparently unrelated behavioural tendencies of which the origin is unclear and the sudden successful combination puzzling. On the other hand, there are many aspects of primate behaviour which do not pose such problems. To apply intentional language to these aspects, to the behaviour of fruitflies (or, for that matter, to the growth of plants), may bring us back to the point where it was decided, and rightly so at that time, that such language can be confusing and meaningless.

References

Griffin, D. (1976). *The Question of Animal Awareness.* New York: Rockefeller University Press.

Köhler, W. (1917). *Intelligenzprüfungen an Menschenaffen.* Reprinted 1973, Berlin: Springer.

Waal, F. de (1982). *Chimpanzee Politics.* London: Cape.

COMMENT

D. PLOOG

The distinction between causal and cognitive explanations of behavior is, I believe, utterly important for our subject. But I cannot see why there is no way for reconciling causal with cognitive explanations. To me, these are two different levels of analysis. If I am to describe an animal which strives for food and demonstrates insight in complex situations (detours, opening of locked boxes, etc.) in order to get access to the incentive, I would not hesitate to call this behavior intentional. On the other hand, if I am interested in the causes of the animal's motivation, I could look for its glucose level, for its temperature or for neurons in the hypothalamus which discharge on the sight of food. This is one of several levels in which causation of specific behavior can be investigated without interfering with the aspect of intentionality.

COMMENT

G. ETTLINGER

Certain theoretical issues may be resolved if the comparison between primate and man is re-specified: primate/human *child.* (No primatologist has claimed capacities for any ape in excess of those of 1–2-year-old children.) The non-linguistic child, like the non-human primates, cannot be expected to give an account of its behaviours (reasons, intentions, etc.). One may then ask how far the behaviour of child/non-human primate is comparable, instead of asking whether the non-human primate has a particular behaviour (for example, language).

However, even then the level of a behaviour must be carefully defined. It is highly probable that the automatic/non-automatic dichotomy is too simplistic, as more complex levels will have evolved gradually. If then we are dealing with multiple levels (with automatic/non-automatic merely representing the extremes), an operational definition for each level (e.g. perception, cognition, intention, etc.) must be sought, and will evolve by experimentation. My personal hunch as regards the issue of how to define 'cognition' would be to emphasize the adaptive aspects – a goal or outcome is achieved, even though circumstances change on repeated elicitations of the behaviour. However, I also believe that ultimately the various levels will be definable by their neurophysiological correlates. Thus, perception could be correlated with neural activity in certain systems, cognition with neural activity in other, so-called 'higher' (i.e. further remote from the input) systems.

REPLY

R. HARRÉ

De Waal seems to have missed the pivotal point of my argument, which was precisely to *deny* the plausibility of applying intentional concepts indiscriminately to all animal behaviour. I warned that the necessity to use GDA forms of description should not tempt us into forgetting that more stringent conditions are necessary for using intentional descriptions. This aside I am in full agreement with the rest of de Waal's remarks.

B. Language and the description of communication systems

6. Ethology and language

EDWIN ARDENER

Reynolds (1975) lists the differing terminologies used by students of social behaviour of the rhesus monkey, *Macaca mulatta*. He shows that the significant units of behaviour are difficult to define, and that it is far from certain that the 'same' inventory of perceived units can be derived from the analyses of various primatologists. In one table, listing four authors' usages (here I label them A, B, C and D) there are 16 possible behaviour units. In another table using three authors only (B, C, and D) a further 8 units emerge. I do not wish to enter into the technical features of this valuable paper, but to draw attention to some of the broader implications of the variations described therein. From the four-author series I give a characteristic selection of behaviour units, described as follows:

Unit	A	B	C	D
1	Hough-hough	Hough	Bark, pant, threat, roar, growl	'!Ho!'
2	Coo coo	Food call	Food call	'Kōō'
3	Eech	Screech	Screech	'ēēē'
8	Looking directly at opponent	Aggressive look	Glare or scandalized expression	Stares at
11	Rigid body posture and stiff legs	Haughty walk	Slow pacing	Holds tail erect

After Reynolds (1975:282, Table 1).

For the three-author series the following is also a characteristic selection:

Unit	B	C	D
1	Feeding hough	Food bark	'!Ho!'
2	Bark	Shrill bark	'Ka!'
3	Splutter	Gecker	'ik, ik, ik . . .'
4	Submissive sit	Cat-like sit	Looks 'apprehensively' (towards)

After Reynolds (1975:283, Table 2).

These examples bring out clearly some immediate problems in the supposedly direct observation of significant behaviour. First, we may note that the human observer's cultural background penetrates even his description of primate behaviour. It is interesting, for example, that a rendering of rhesus vocalization is attempted in several cases. Authors A and B characterize the monkey sounds by approximations from English folk-phonetics. Author D appears at first sight to introduce a more 'international' standard, but it seems on a closer examination that the '-ōō' and 'ēēē' notations are phonetically [u:] and [i:] or the like, and not, for example [o:] and [e:]. Thus D introduces slightly more graphical confusion than A and B. The impression given by the choices of all three authors is of a conscious, almost literary exoticism. The renderings of authors A and B relegate the rhesus sound to a region of English graphemics which is a by-word for the lack of consistent phonetic reference. Their *hough* (or *hough-hough*) is no doubt felt to be like a *cough* [kɔf], a *hiccough* [hikʌp] or a *huff* [hʌf] rather than (say) like *bough* [bau], or *through* [θru]. There is the further cultural association of Irish words like *lough* [lɔx] (or anglicized [lʌf]). The obsolete archaic spelling *hough* (apparently [hɔk]) for *haw* (as in 'hem and haw') or possibly for *hawk* (as in 'hawking and spitting') also exists, although it is uncertain whether the reminiscence is deliberate. We may note that *haw* and *hawk* in these usages themselves occur in restricted, fairly stereotyped, collocations, and are for some speakers nearly obsolete. Further, we may note that for author C the articulation is represented variously by *bark, pant, roar* and *growl*, none of which is normally regarded as synonymous with the others let alone with *haw* or *hawk*.

For author D (who may be from an American milieu) '!Ho!', with its idiosyncratic exclamation marks front and back, presents an exoticism of a different kind, as does his use of both diacritics and repetition to represent length in other examples. Written English contains among its stylistic conventions a loose but not totally open set for the rendering of the outlandish and uncouth. When speakers of a language attempt to represent non-human, or totally alien sounds, they commonly do their linguistic neighbours (whether speakers of non-standard dialects or foreigners) the honour of acting as models. Such neighbours are like 'nature' rather than 'culture' as Lévi-Strauss would point out (see also Leach 1964). There is certainly an archaic, dialectal or Celtic hint about the descriptions of rhesus monkey vocalizations by authors A and B, and a primitive or faintly folk 'Red Indian' flavour about author D's '!Ho!'. For a hint from another cultural background, Jorge Luis Borges, Argentinian, in his *Ficciones,* invents an outlandish country which he calls 'Tlön'. Surely only a speaker of Spanish would chance upon this only apparently 'random' choice, containing as it does 'tl' from the exotic Amerindian-derived area of the Castillian lexicon – found mainly in Mexican topography and nomenclature – and 'ö' from the European north – alien in orthography, sound, and cultural association.

When we confront the attempt to label rhesus utterances, we may well argue that cultural peculiarities are at their most evident when humans try to represent the *absence* of cultural characteristics. If, as Evans-Pritchard (1940) suggested long ago, we define ourselves by opposition to others, then 'the other' is not an open category of infinite possibilities, but is in turn defined by its opposition to ourselves. Each culture inevitably generates its own perception of what is, either as dream or nightmare, its 'other'. We might be tempted to ask in the particular case, why, since a phonetic approximation was being offered, all the well-developed resources of the International Phonetic Alphabet (IPA) were not brought into play. A Japanese scientist would, at least, then not need to know English in order to understand the representation of a rhesus utterance. The use of the IPA would, however, only confuse the matter further, for rhesus phonetics are not human phonetics. Evidently the only adequate solution would be a special rhesus phonetic inventory. Even when that solution was reached, the labour might be of only academic interest since rhesus utterances are no doubt so stereotyped that a numbered inventory of sample tape-recordings might be more convenient. Any particular written form would then be of merely mnemonic function. In that case whether we write *hough-hough, hough, '!Ho!'* or *No. 1* is of no particular moment, thus returning us full circle.

There is, however, a difference in consciousness between our conclusion and that demonstrated by primatological practice. For we see that even when Reynolds' authors abandon a quasi-phonetic notation of articulations for a 'description' of behaviour, they draw the same problems with them into this sphere. We meet here *geckering,* a term from the world of English dialect. *Gecker* is meaningless to most modern southern English-speakers – its basic usage seems to derive from 'a gesture of ridicule' (see *Oxford English Dictionary* s.v. *geck*). I have furthermore heard a primatologist state, no doubt in an unthinking moment, that the 'correct' word for one kind of rhesus grimace was *girning*. *Girn* derives from *grin* by metathesis in south-western English dialects (Wright 1913:134). It is known to standard English speakers (when it is known at all) through the rural folk-custom of holding *girning* competitions, to discover who can pull the ugliest face while looking through the frame of a horse-collar. Nothing could be more culturally specific than the grotesque grimaces of *girning*. Its dialect origin no doubt lends it the association with the outlandish required to describe monkey behaviour.

We may press the matter further. For one unit of behaviour, author C chooses *gecker*, D uses *ik, ik, ik,* but B selects *splutter*. It may be thought that *splutter* at least is a neutral descriptive term, but this belongs to a domain of lexicon which ultimately is as culturally idiosyncratic as *gecker*. For example, translating terms of this sort between languages can be particularly difficult. Informants state that only this, or only that, is a 'splutter', perhaps with obscure demonstrations – and with careful exegesis on the affective implications without which (or with which) it is not (or is), truly, a splutter. The ethologist's problem is that demeanour is rule-governed in human societies (Ardener 1973). The natural language terms available to cover animal sounds and acoustic gestures are likely to come dressed in highly emotive cultural associations. Animal behaviour is likely to recall uncontrolled, childish or ill-mannered demeanour, and the terms used bring semantic overtones into the supposedly neutral description.[1]

When we come, therefore, to more elaborate delineations of rhesus activity, as with author C's 'glare or scandalized expression' or B's 'haughty walk' we are in the thick of the ethological problem. If 'haughty walk' is acceptable could we also accept 'sagacious nod' or 'admiring stare' as possible rhesus behaviours? These remarks are not intended to make jests at the expense of ethological description. All the problems of scientific investigation are present in these examples. Ethologists are increasingly aware of these questions (cf. a recent unpublished criticism by Chalmers of the category of 'play'). For a discipline which wishes to determine objectively what animals and humans have in common, it is an issue of critical importance if natural language has already prejudged the issue by 'contaminating' the descriptive instruments with evidence of humanity (see also Callan 1970, and Crick 1976:100–8, for further discussion). For social anthropologists the important conclusion, always worth repeating, is that the observation and the labelling of 'behaviour' are inseparable from the importation of socially derived meanings – even when observing monkeys.

In putting forward this preliminary note (prepared originally for a different purpose) on the highly cultural (and not merely 'human') features that can colour our examination of primatological matters, I do not prejudge the wider question of whether primate communication occurs and can be studied. The use of so-called sign-languages in dealing with primates is, however, only one area in which many misconceptions arise, which are part of the problem lightly sketched in by this contribution.

Note

1. For the development of the term 'behaviour' in scientific usage see Ardener (1973). The term always meant 'socially ordered activity'. Its appearance in scientific use in the 1850s represented a demonstration that nature *was* orderly. 'Behaviour' was subsequently also used to denote an activity for which the demonstration of orderliness was only an aspiration – thus the paradox of 'random behaviour', and the desocialization of the term.

References

Ardener, E. (1973). Behaviour: a social anthropological criticism. *Journal of the Anthropological Society of Oxford*, **4**, 152–4.

Callan, H. (1970). *Ethology and Society: Towards an Anthropological View.* Oxford: Clarendon Press.

Crick, M. (1976). *Explorations in Language and Meaning: Towards a Semantic Anthropology.* London: Dent (Malaby), (New York: Halstead).

Evans-Pritchard, E. E. (1940). *The Nuer.* Oxford: Clarendon Press.

Reynolds, V. (1975). Problems of non-comparability of behaviour catalogues in single species of primates. In *Contemporary Primatology*, ed. S. Kondo, M. Kawai & A. Ehara, pp. 280–6. Basel: Karger.

7. Must monkeys mean?

ROY HARRIS

There are two conceptual muddles about meaning which persistently obtrude into debates on the subject of animal communication. They may be identified by reference to the following – not untypical – example of human rationalizing about the meaning of animal signals. It comes from a report in *The Times* (22.12.80) on the work of Seyfarth and his colleagues on the 'alarm' calls of the vervet monkey.

The report describes the calls in question in the following way:

> The monkeys run up trees on hearing the leopard alarm, look up and flee into dense bush for the eagle alarm, and stare at the ground around them for the snake alarm.

The investigators, the report says,

> began by listening to and analysing the calls, and noting the circumstances in which each type of call was used and what the monkeys did when they heard them.
>
> Although it soon became clear that the monkeys took appropriate action on hearing each call, that was not enough to prove that the monkeys really knew what each call meant. It was still possible that the calls had only alerted them to danger and that they had looked round to see what the danger was.
>
> The team proved that the calls had a specific meaning by playing tape-recorded calls from a hidden loudspeaker when no predator was present; in each case the monkeys took the evasive action suitable for the specific predator signalled by the call.

The reader of this *Times* report is in effect being invited to accept the following argument. If our information about vervet monkeys were restricted to observing but not interfering with their natural behaviour, we should never know for sure whether they really knew the meaning of the signals they appear to use, or whether we were just reading that interpretation into our observation. For, in natural circumstances, a signal does not occur except concomitantly with the relevant danger. But by divorcing the signal from the danger experimentally, the human investigator is able to prove that the monkey does not merely behave *as if* he knew what the signal meant, but that he really *does* know.

Whatever that example tells us about animal communication, it certainly tells us a great deal about human thinking about animal communication. And before we can make much progress with elucidating the meaning of primate signals, we need to elucidate what human statements about the meaning of primate signals mean.

The two conceptual muddles this paper is concerned with can be thought of as two possible lines of objection to the way the notion of meaning is invoked in the *Times* report. One line of objection is to say: the Seyfarth team's experiment is far from proving that the monkeys knew what the signals meant, for it does not even demonstrate that the calls were really part of a signalling system in the first place. The other line of objection is to say: even if the calls *were* signals, the experiment supplies no proof that the monkeys knew what they meant, but merely that they reacted in specific ways.

Both objections are characteristic of what may be termed the 'apartheid' position on communication. The 'apartheid' thesis is that human communication and animal communication are equally valid, but separate; or at least must be treated as separate under any respectably 'scientific' approach. For otherwise, allegedly, the danger is that by using terms like *meaning* indiscriminately of both human and animal communication we shall allow our terminology to prejudge for us certain very important questions about the interpretation of animal behaviour and its comparability with human behaviour. In particular, we may be led to suppose that animals have certain mental capacities which they in fact lack. For 'meaning' is a notion inextricably bound up with human mastery of language. But if language is qualitatively different from all forms of animal communication then, so the argument runs, it is at best metaphorical and at worst totally misleading to apply to animal signalling the conceptual framework appropriate to linguistic signalling.

There are a number of different strands tied up already here. Some of them there will be no time to untangle in this chapter: they will simply have to be cut through. Of these, the main one is the implication that by somehow developing a scientifically fastidious aseptic terminology we could actually investigate the processes of animal communication in a kind of conceptual vacuum, where no germ-laden assumptions borrowed from human analogues could intrude. Perhaps no more need be said about this than to point out that if indeed one were to carry out such a programme rigorously, the first casualty would have to be the concept of 'communication'; and the trouble with that is that at one stroke it abolishes the field of study altogether. If we are unwillingly to be such ruthless abolitionists, then we might as well address directly the question of why it might be objectionable to include 'meaning' in our conceptual apparatus for analysing animal communication.

There are two main reasons usually given for reluctance to equate the meaning of animal signals with the meaning of linguistic signals. One is that animal signals are characteristically 'stimulus-bound'. The other is that most animal signals are restricted to communicating one of a limited, nonproductive set of fixed messages known to all members of the species. The meaning of a linguistic signal, by contrast, is alleged to be typically free from stimulus control, and is not necessarily drawn from a set of messages known in advance to all human beings. In short, the nearest human equivalents to animal signalling are regarded as being certain forms of spontaneous, involuntary expression, such as blushing, weeping, inarticulate exclamations of pain, etc. Although these forms of expression are in one sense communicative, they are of an entirely different nature from linguistic communication. Or so, at least, the 'apartheid' theorist holds.

The 'apartheid' position has a long and respectable intellectual ancestry, going back in one form or another to Classical antiquity. It was emphatically reaffirmed in the nineteenth century by those opposed to Darwin's disturbing claim that there is no fundamental difference between man and the higher animals in respect of their mental faculties. 'Language', said Max Müller in 1861, 'is our Rubicon, and no brute will dare to cross it'. But Müller's Rubicon metaphor was a curious one. Daring to cross Rubicons is a question of volition, courage and expediency. They only dare – or do not dare – who can. Whereas the more general view about brutes is that they cannot achieve language, even if they would. Bertrand Russell observed that however eloquently a dog may bark, he cannot tell you that his parents were poor but honest. Frege apparently thought that even if a dog could bark numerals, he would still be incapable of counting, because a dog would be unable to grasp any concept of number.

On the Fregean view, even if a dog were successfully trained to engage in barking performances which resembled counting – and there have been circus acts of this kind – that would prove nothing. Whatever type of counting performance a dog put on, he could not possibly 'mean it' (i.e. the animal could not be counting in the sense that human beings count). Even if a dog consistently barks once when shown one bone, twice when shown two bones, thrice when shown three bones, and so on, nevertheless, by barking x times the dog cannot really 'mean' – nor can his barking 'mean' – that there are x bones.

Now perhaps monkeys are cleverer than dogs. (For present purposes let us use the term 'monkey' in accordance with the broadest definition given by the *Shorter Oxford English Dictionary*, i.e. as covering 'any species of the group of mammals closely allied to and resembling man'.) And perhaps because an animal is closer to man than a dog is, we might be more favourably inclined to treat examples of its quasi-counting behaviour as instances of genuine counting. Although overtly suspicious of Fido and his circus trainer, we may be secretly impressed by the sight of a chimpanzee pressing buttons in a university laboratory. Indeed, if the warnings of the 'apartheid' theorist are to be heeded, then primatologists more than most should be on their guard against deliberately or unwittingly assimilating animal signalling behaviour to human signalling behaviour.

In short, even a whole troupe of talking monkeys will not satisfy the 'apartheid' theorist in his most intransigent mood. He will still demand proof that, as Descartes once put it, 'they are thinking about what they say'. Otherwise, he will see no ground for supposing that they know what they mean.

Two reactions to this neo-Cartesian attitude are of interest here. One is an attempt to turn the tables on the 'apartheid' theorist and force him to grant that if it can be shown that primates are after all capable of mastering signalling systems of a certain degree of complexity, this shows that primates are, at least to that degree, capable of thought. A quite opposite reaction is simply to concede the argument to the 'apartheid' theorist, and disclaim any pretension that using terms like 'meaning' in connection with animal behaviour implies that animals communicate with one another in anything like the sense that human beings do.

Both reactions are equally misguided, although in quite different ways. Let us first of all focus upon the more widespread, which is probably the second. What is at issue here can perhaps be brought out more sharply if we take a non-primatological example where there is no temptation to treat animal signalling as a rudimentary form of language. The following is the text of a recent advertisement for a mosquito repellent.

STOP MOSQUITOES BITING!

With the incredible —

A tiny transmitter, the size of a throw-away lighter, that repels biting mosquitoes indoors, or out, and with a range of 20 feet.

Here's how it works.

It's a well proven scientific fact that the only mosquitoes that bite humans are mated females needing blood to develop their eggs.

All other kinds feed off plants.

— transmits a tiny high-pitched hum almost inaudible to the human ear.

But not inaudible to the female mosquito.

She thinks she hears the sound of a male mosquito. And that's the last sound she wants to hear, because she's mated already.

So she stays away.

Incredible?

Yes, but — works amazingly well. Its replaceable 1.5 volt battery (not included) will give you up to one year's continuous use.

At £5.00 — is a must for fishermen, campers, caravanners, bird-watchers, barbecuers — anyone who normally has to put up with the curse of mosquito bites.

Try it. We're so confident that you'll find — indispensable, that we're making an unconditional 1 year money-back offer. If you aren't absolutely satisfied with your — send it back to us and we'll refund your money in full.

The stance we are just here concerned to identify is one which would maintain that to speak of the 'meaning' of the mosquito signal is unobjectionable *provided that* we take care to purge the meaning-assignment of any implications which would assimilate it to the kind of meaning-assignment typically employed in the case of linguistic signals. This can be done, it is urged, by restricting the meaning-assignment to just one of (at least) three possible types of statement.

The three types of statement in question may be distinguished as follows:

(*a*) stating what the meaning is by supplying an equivalent linguistic expression ('signal *s* means "*m*"'),
(*b*) stating what the meaning is by giving some speech-act classification ('signal *s* is a *p*-type act'), and
(*c*) stating that the signal conveys certain types of information about the signaller or the situation ('signal *s* means that the signaller – or the situation – is *q*').

Statements of the first type are conceded to be objectionable because there is no viable translation from animal signals into sentences. Statements of the second type are conceded to be objectionable because speech-act classifications are borrowed from human interaction. Whereas statements of the third type are held to be admissible, provided the information conveyed is in some sense physical or physiological information, as opposed to culturally mediated information.

That is to say, provided the meaning-assignment does not fall into the categories (*a*) (e.g. 'the mosquito signal means "Beware", or "I'm coming to get you"') or (*b*) (e.g. 'the mosquito signal is a warning'), but only into some appropriate section of category (*c*) (e.g. 'the mosquito signal means that there is a male mosquito present'), then it is harmless and perfectly correct to speak of the signal as having a meaning. For this can in no way be taken to imply that the participants are mini-human-beings.

The mosquito repellent advertisement is an interesting example precisely because the text of this advertisement itself gives the lie to the latter claim quite directly. The fact is that even the most neutral imaginable meaning-assignment of category (*c*) never occurs except as part of an explanatory account of communicative behaviour which is basically of the same kind as we give for human communicative behaviour. For this is a quite general function of meaning-assignments. It would be a mistake here to try to appeal to different senses of the verb 'mean' or the noun 'meaning'. We cannot keep out of trouble – if trouble there is – by trying to dissociate statements about what the signal means from statements about what the creature means, or what its giving or receiving the signal means. All these in the end are part of the same kind of explanatory account.

In the example just quoted, the behaviour of the female mosquito is explained by reference to the meaning of the signal for her, and a reason is then given why, interpreting the signal as she does, she takes evasive action. For all its unsophisticated lay presentation, this is perfectly typical of a certain general pattern of elucidation of animal behaviour which is widely encountered in more academically respectable contexts, and it brings out rather clearly the function of meaning-assignments referred to above. A basic requirement of such assignments is to provide a plausible explanation of some type of behaviour associated with the signal, either in general or in particular contexts.

In such cases, the position of the analyst in animal communication is quite similar to the position of that hypothetical linguist sometimes invoked by philosophers, who supposedly has to infer from his observation of the natives' behaviour in the jungle what meanings to assign to the noises they utter. Thus the reason why this hypothetical linguist puts down 'rabbit' as the meaning of the native utterance *gavagai* (rather than, say, 'what time is it?' or 'I wonder what's for dinner') is implicitly taken to be that it provides a better explanatory link between certain types of native behaviour. The relation between rabbit-spotting activities and non-rabbit-spotting activities is better accounted for on the assumption that *gavagai* means 'rabbit', or 'look at that rabbit', or 'what a splendid rabbit', or something similar. But any such hypothesis about *gavagai*, however we frame it, involves certain important presuppositions about the normal conduct of interactions involving signals of that kind.

The presuppositions in the linguistic case might be summarized as follows: that linguistic activity is part of a rational pattern of behaviour of the individuals concerned. It is interesting to consider how comparable presuppositions are also latent in accounts of animal communication. The female mosquito of our advertisement, for instance, is manifestly represented as a rational mosquito. She is already mated: *therefore* she does not wish to be troubled by further attention from male mosquitoes: *therefore* she takes evasive action on hearing a sound she takes to signal the presence of a male mosquito. This explanation clearly adds considerably to the plausibility of the advertisement. It gives the potential purchaser an extra reason for believing in the probable effectiveness of the device he is being invited to buy. The advertisement would be far less convincing if it merely claimed that observation showed female mosquitoes to have a tendency to avoid coming within 20 feet of the source of a certain type of high-pitched vibration. In order for that kind of statement to be part of a plausible explanation of the effectiveness of the device on sale, it would need to be supplemented by an account of the biological mechanisms underlying the female mosquito's avoidance of the vibrations in question – a much more complex and difficult matter. Thus postulating that the female mosquito interprets the sound as a signal, and gives it a meaning, makes available a mode of explanation which does not need recourse to biomechanical analyses at all.

Precisely for that reason, however, the validity of the rationale lays itself open to challenge by the 'apartheid' theorist. It strictly makes no sense, he argues, to say that the female mosquito takes evasive action because she mistakenly interprets the transmitter sound as signalling the presence of a male mosquito, and therefore, being already mated, seeks to avoid a possible encounter. This is just Toytown language. It attributes inferences and propositional knowledge to a creature manifestly incapable of either. Thus not only is it literal nonsense, but it is not even justifiable metaphor. It merely obscures our scientific understanding of the reasons for the female mosquito's behaviour.

The force of that objection may be more obvious in the case of a mosquito than in the case of a monkey. But for some that only makes the adoption of Toytown language all the more objectionable in respect of primate behaviour, and the desirability of avoiding it greater.

The point to realize here is that, in spite of appearances, there just is no contest between the devotees of Toytown language and those who would ban it. For on the one hand Toytown language cannot somehow be justified by blandly disclaiming what is in effect its principal explanatory role. But nor, on the other hand, can Toytown language be ruled out of court on such curious grounds as that mosquitoes cannot 'really' reason. That would be like trying to ban advertisements selling pills for constipation on the grounds that the pills themselves are neither for nor against constipation. To anyone who urged banning such advertisements *for that reason,* one would simply have to try to point out that the mistake lay not in the language of the advertisement, but in a naive or perverse interpretation of it.

Turning from mosquitoes to monkeys, we find the whole argument about the success or failure of programmes to teach primates the rudiments of language shot through with question-begging Cartesian presuppositions. On the one side it is assumed that if monkeys demonstrate an ability to use signs effectively in interaction with their trainers that shows they 'understand' the concepts associated with the corresponding word-uses in English. On the opposite side, it is claimed that such programmes fail because the performance of the trained monkey does not demonstrate conclusively that it really knows what the signs or sign-combinations mean. This type of debate is as absurd on the one side as it is on the other. It presupposes that language is by definition, as Malinowski once put it, simply 'the countersign of thought'. More exactly, it presupposes the validity of a Saussurean bi-planar model of language.

In the most ludicrous cases the debate goes something like this. On the one side the fact that the monkey can use a certain gesture systematically in trainer interactions to elicit the reward of pieces of apple is glossed as 'the monkey knows that this sign means "apple"'. On the other side the validity of this gloss is denied on the ground that all the monkey has shown is that it has found out how to get its trainer to give it pieces of apple, and merely being able to do that fails to demonstrate that the monkey knows that the sign has a meaning at all. This a like a duel fought out with fairground firearms, of which the one thing you can be sure is that when the target is lined up perfectly in the sights, then the shot is bound to miss.

Or take the argument about whether or not apes can use expletives. On the one side it is claimed that when the animal uses the Ameslan sign for 'dirty' in contexts where the thing in question is perhaps objectionable but not literally unclean it is giving vent to the same kind of disapproval or annoyance as when human beings say things like 'it's a dirty swindle'. On the opposite side it is claimed that these misuses of the 'dirty' sign are not true expletives, because they can be accounted for on the assumption that the animal knows that signing 'dirty' is a reliable way of getting out of a situation it dislikes and being sent to the toilet.

Many of these *dialogues de sourds* in the ape-language controversy could never have taken place if the controversialists had learnt the basic lessons about meaning and communication which were spelled out by Wittgenstein in the *Philosophische Untersuchungen*. However that may be, the problem for any neo-Cartesian approach to the question of comparing human and animal communication remains what it always was. How do you prove or disprove, without begging the question, that certain forms of behavioural response are manifestations of certain forms of conceptual thought? Two extreme stances may be contrasted here. One is to insist for instance in the case of number concepts, that there is just no sense to be made of the claim that an animal can count, unless the animal has mastered some system of arbitrary vocal or other signs, and can correlate them convincingly with certain numerical differences. (This is a view which links up with the second muddle about meaning, to be considered in the second part of this paper.) The opposite stance would reject this claim, on the ground that it simply makes mastery of signs a prerequisite of tests for concepts, and is hence in effect circular. All one can do, according to this opposite view, is to determine whether an animal does or does not respond differentially to systematic differences of number; for in terms of the behaviour natural to the animal that is all that counting could mean.

Now whether we are dealing with number concepts or concepts of any other kind, the same conflict of criteria arises. Hence it would be a mistake to argue from the premise that some animals, notably chimpanzees, just *have* shown themselves able to classify consistently by means of arbitrary signs. And it would be a mistake for at least two reasons. One is that what Washoe can do does not automatically provide a yardstick applicable throughout the entire domain of animal communication, or even primate communication. Even if Washoe could be taught to do multiplication by logarithms to everyone's satisfaction, that would still leave the conflict of criteria unresolved. But the second and more basic reason is that there is no classification-performance by a chimpanzee which cannot be explained away in principle on assumptions other than that the animal knows the meanings of the signs used. It is this which is significant, and reflects a central weakness in the Cartesian position. What happens is that either the neo-Cartesian principle 'no language without thought' gets translated experimentally into terms which make it impossible to detect thought in the absence of linguistic signs or their equivalents – but nonetheless fails to resolve the question of how the signs have to be used in order to prove the presence of thought – or else the notion of 'thought' is etiolated to a point where almost any kind of systematic response will count as conceptual discrimination at the relevant level. With the result that – to borrow an example from Peter Geach – there is nowhere to stop short of conceding that even earthworms understand the concept of angular magnitude, since experiments show that when given a triangular piece of paper they will choose to pull it into their burrow by the sharpest angle.

In neither case can much progress be made with the problem of the meaning of animal signals. All that happens is that analysis is sidetracked into an irrelevant debate about the possibility of drawing a division across what is no doubt a continuum of communicative behaviour. Again the fight has to be declared a 'no contest', because what appeared to be a substantive issue turns out on investigation to be logomachy.

Two brief appendices to the above remarks on the first of our two muddles about meaning are in order at this point. 'Appendix A' concerns the difference between stimulus-responses and speech-acts.

There are those who ask (as a rhetorical question): 'Could a monkey possibly have Gricean intentions?' (Could a monkey intend the performance of *x* to produce an effect in an audience by means of the recognition of this intention?) If the answer, as expected, is negative, then whatever else a monkey might be doing, it could not possibly be meaning anything by what it does (in the sense in which this is dependent upon the existence of a non-natural meaning convention, as in human signalling systems, including languages).

The reason for asking the question is that it is supposed that unless we can somehow bring intentions into the act, we are left with unadulterated behaviourism, and behaviourism is a Bad Thing. In particular, it is a Bad Thing because it fails to distinguish meanings from outcomes or reactions.

Unfortunately, the rhetorical question in these cases rebounds on the questioner. Imagine a human experimenter, Smith, who embarks on an experimental programme to see if monkeys can be taught the rudiments of linguistic communication. Smith is a Gricean, in that he conceives his task as that of getting monkeys to produce an effect in an audience (viz. himself) by means of the recognition of this intention. The difficulty is that however many times the monkey selects the correct token or presses the correct button for his banana, Smith cannot decide whether he has spotted a Gricean intention there or not. It is doubtless no consolation to Smith to point out that, *mutatis mutandis*, the monkey would have exactly the same problem: it would never be able to decide whether Smith was acting on the basis of Gricean intentions or not. Smith and the monkey could carry on in this state of uncertainty indefinitely, without it ever affecting the communicational efficiency of the system they have established for negotiating over bananas.

'Appendix B' concerns the importance assigned in assessments of language-learning programmes for monkeys to detection of the 'Clever Hans effect' (so named after a celebrated circus horse, whose feats of understanding are supposedly explained by his aptitude for picking up cueing by trainers and experimenters). The significant thing about this is the assumption that an animal is somehow 'cheating', or a test badly designed, if the 'Clever Hans effect' is not eliminated.

Nothing could illustrate more clearly the neo-Cartesian presupposition that what has to be established is grasp of an abstract correlation between sign and concept, unmediated by any form of participant interaction. In other words, the ultimate criterion of language is tacitly taken to be the fact that man is capable intellectually of decontextualizing his own verbal behaviour. It is rather like believing that the supreme test of ability to drive a car is being able to answer questions about the Highway Code. There is a confusion, in other words, between two issues: what constitutes communicational mastery of a sign system, and what constitutes proof of the ability to perform certain intellectual abstractions based upon mastery of a sign system.

The second muddle about meaning arises from a variant version of the 'apartheid' thesis, which attaches less importance to what the animal can do, but more importance to what the signalling system can do. This is a position which has proved attractive to a good many linguists.

For example, Fromkin & Rodman cite the following evidence to show that although there is no doubt that honeybees can communicate messages about the location of sources of nectar, they do not communicate this information *linguistically*. The evidence comes from an experiment in which, instead of being allowed to fly to the source of nectar, the bee was made to walk. Upon returning to the hive to communicate the whereabouts of the nectar to his fellows, the ambulatory bee misinformed them, having overestimated the distance of the nectar by a factor of 25. This shows, say Fromkin & Rodman (1978, p. 42), that the bee when forced to walk 'had no way of communicating the special circumstances or taking them into account in its message. This absence of *creativity* makes the bees' dance qualitatively different from human language.'

Here the Scylla of arguing about whether the bee 'thinks' is avoided only by falling foul of the Charybdis of arguing about whether the bee's communication system is a 'language'. In the event, the argument produced is worthless. It is rather like saying that if you blindfold a speaker of English and discover that he can no longer make reliable judgments about which is north and which is south, that proves that his earlier statements 'this direction is north' and 'this direction is south' were not acts of linguistic communication at all.

Again, there are various threads knotted up in this second muddle. They could with more time be untangled, but again some of them must here just be snipped through. One such thread is the criterion invoked by Hockett, who dismisses the honeybees' communication system because it fails – unlike languages – to exhibit duality of structure. The difficulty with this criterion is that very much depends on what one is willing to count as duality of structure. Ultimately, it seems, there is no communication system, however simple mathematically, which cannot be analysed as having at least two levels of structure (Harris 1980, pp. 25–8). So invoking duality of structure as a criterion for language turns out to sink the very ship it was intended to launch.

Another such thread is the reluctance to treat animal signals as meaningful in a fully-fledged sense of the term because of the misapplication of a doctrine from information theory. It goes like this: if animal signals are just responses to stimuli, then they must be predictable even if investigators have not yet succeeded in working out the exact rules of prediction. Whatever is fully predictable is, from an information-theoretic point of view, meaningless. Meaningfulness, it is held, implies choice. Since the occurrence of a signal which is automatically triggered by some situational factor precludes choice, such a signal cannot be meaningful.

Although this doctrine plainly fails to square with common sense (for the fact that a fire alarm is automatically triggered by a certain rise in temperature does not make the alarm signal meaningless) it is a doctrine which has not been without influence in linguistics. Linguists have claimed, for instance, that 'an utterance has meaning only if its occurrence is not completely determined by its context' (Lyons 1968, p. 413). Hence an utterance like 'How do you do?' is said to have no meaning if it is mandatory in certain social circumstances. In such a case, it is held, all the relevant semantic facts about the utterance are included in a mere statement to the effect that it is used obligatorily in the contexts in question. It is futile, allegedly, to insist that it must 'mean' something over and above its 'use' (Lyons 1968, p. 414).

It is not difficult to see that the reasoning, when made explicit in this way, is highly dubious. There is an equivocation over the notions of 'determination' and 'choice'. For meaningfulness implies choice only in the limited sense that to ask what someone meant (communicationally) by doing *x* normally presupposes that he could have chosen not to do *x*, or to do *y* instead. But depriving an individual of choice in a communication situation does not *ipso facto* render what he does communicationally meaningless, any more than leaving a chess player with only one move he can possibly make renders that move invalid. In rituals, for example, everything that is done by the participants may in one sense be totally predictable, and the scope for individual choice as to what to do next may be nil. But it would be patently absurd to draw the conclusion that no particular element in a given ritual has any meaning. It would make no difference either whether the participants in a given ritual always acted under hypnosis, or had unwittingly been preconditioned to respond to certain subliminal cues. Analogously, in the case of primate interactions, the extent to which particular behaviours are learned, innate, or stimulus-conditioned are quite separate questions, which must not be confused with that of the meaning of the signals involved.

Primatologists should take the honeybee example as a warning. It provides an object lesson as to how the argument about the meaning of primate signals is likely to go if anyone insists on squeezing it into the Procrustean bed borrowed from linguistics. What will happen is that questions of meaning will be referred to assessment in the light of the signalling systems to which the signals belong. If the signalling system is judged not to be comparable to the language systems of *Homo sapiens* in certain crucial respects, this will be taken as a ground for denying meaning to the signals in question. In short, meaning will be taken as emanating from the communication system, rather than from users of the system, whether human or non-human. That is precisely why the question of whether chimpanzees can be taught to operate a sign system equivalent to English in certain crucial respects has loomed large of late in linguistics. Being able to use a communication system of the right type is implicitly taken as decisive for resolving the issue one way or the other. No further proof is required, and no other evidence is regarded as relevant.

This alternative to the Cartesian position is equally unsatisfactory. Just as the principle 'no language without thought' leaves us with the problem of deciding what are the minimum requirements for thought, the alternative 'no language without a language-system' leaves us with the problem of deciding what are the minimum requirements for a language-system. But it is difficult to see how this can be other than a matter of stipulative definition. Again, the question of the meaning of primate signals is sidetracked by a quite irrelevant debate. There is no substantive issue about meaning, and the muddle essentially consists in believing that there *is* one.

The source of both muddles discussed in this chapter is the same. It is failure to recognize that statements about meaning have a role to play in communicational analyses of interactions which cannot be usurped by any other concept. Communicational analyses and explanations constitute a type of account which is entirely *sui generis*. Such accounts are not reducible to or equatable with biomechanical accounts or mentalistic accounts. Nor, thirdly, are they accounts which are valid only if the communicational signals in question meet certain arbitrarily fixed criteria of structural complexity or practical efficiency. What such signals can or cannot do is just as irrelevant as whether the agents who operate them are animate beings or robots.

The answer to the question 'Must monkeys mean?' is: 'If any communicational account is to be given of their interactional behaviour at all, yes they must.' To try to hedge one's bets on whether the observed communicational behaviour of primates allows us to talk about the meaning of primate signals would simply be to opt for a spurious show of 'scientific' caution, but at the expense of logical coherence.

To summarize thus far, the problem with the meaning of primate signals is not – as some people appear to assume – that we have no access to what, if anything, is going on in animals' minds. That notion involves a misconception not only about animal communication but also a misconception about human verbal communication (Harris 1981). In fact, it is simply a version of the Lockean fallacy which equates communication with telementation or thought-transference. We delude ourselves if we think that the descriptive semantics of English is some kind of high-level abstract report about what is going on in the individual minds of speakers of English, or even some hypothesized ideal speaker of English. (That is just the fashionable Chomskyan perpetuation of Locke's mistake.) The semantics of English is – and always has been – part of a human explanatory account of the communicational function of words in a certain form of public social practice which is called 'speaking English'. Similarly, the descriptive semantics of primate signals when we get it, will essentially be part of a human explanatory account of certain social patterns of primate behaviour. To give such an account is already a complex enough task, without dragging in mentalistic red herrings.

Nor is the problem with the meaning of primate signals the fact that animal communication systems are so restricted in comparison with human languages that it becomes misleading to apply the same kind of descriptive terminology to both. The point is well made by Asquith (this volume) that we have no plausible alternative but to use an anthropomorphic conceptual framework in our analyses of animal communication. (That is not, incidentally, necessarily to agree with Asquith's attempt to construe the application of communicational vocabulary to animal signalling as being metaphorical.) We cannot somehow avoid the risks of anthropomorphism – whatever they may be – by trying to talk about primate signals in a terminology which draws no implicit comparison between human and animal communication. Any such attempt must be self-defeating. Again, the task of describing what primate signals mean is already difficult enough, without depriving ourselves of the most useful conceptual tools we have for the purpose.

But where exactly does all that leave us? Is there any way comparisons with human language can actually help enquiry into animal communication, rather than provide potential pitfalls? I should like to suggest that there is. It bears not on such questions as *whether or not* monkeys mean: for clearly they do. But it does bear on the question of *how* they mean. And this is really the question that makes all the difference to constructing a semantics of primate signalling.

The problem of the meaning of primate signals has nothing to do with either the limitations of monkeys' mental capacities or the limitations of monkeys' communication systems. But it may have a lot to do with the fact that human investigators of animal communication could well be looking in the wrong place for clues as to *how* the semantics of primate signalling works. It is possible – and in an evolutionary perspective seems highly likely – that primate communication depends primarily on signalling techniques which human beings have abandoned in favour of verbal signalling. Whereas when human investigators study primate communication, what they often look for are primitive vocal or gestural analogues of verbal signalling. And since these analogues are rather few and far between they conclude that primate communication is a kind of desperately impoverished version of human communication. Now the reason why this difference obtains is very likely to be that monkeys have developed a way of life which puts very little evolutionary pressure on them to evolve a system of communication of the verbal type. In fact, a system of the verbal type would probably serve the purposes of most primate interaction very poorly. Human beings like to think of language as some kind of supreme feat of communicational engineering. But in some respects it is extremely crude, clumsy and inflexible.

This inflexibility has to do with a central feature of all verbal systems of communication: namely, their dependence on linearly discrete signalling – that is, type-token signalling. Hence for human beings it seems obvious that either *The cat sat on the mat* is an English sentence, or else it isn't. There is no way it could sometimes be meaningful and sometimes not, at least within the confines of one and the same communicational system. But this property is neither logically nor psychologically necessary for communication. It seems very likely that in the course of evolution there have been circumstances which favoured the development of non-discrete systems. Also we have to reckon with the possibility that some species may have developed mixed or intermediate systems, which have both discrete and non-discrete features in varying proportions.

A simple example from human communication of the difference between discrete and non-discrete signalling would be the difference familiar to all motorists between the way traffic lights convey messages and the way the motor horn does. The semantics of motor-horn signalling is not amenable to any informative analysis on a simple type-token basis. It allows too much scope for individual variation and is too highly context-dependent. It would be impossible to maintain, for instance, that a single hoot always means one thing and two hoots means something else. So anyone who was misguided enough to set about trying to analyse motor-horn semantics on the assumption that he was dealing with a primitive kind of Morse code would simply be wasting his time. Nevertheless, it is a system of signalling which allows the expression of a wide range of messages, with a considerable degree of subtlety and adaptability. By contrast, the standard system of red and green traffic lights used at road junctions represents communication of the typically verbal type: the semantics of the traffic lights, unlike the semantics of the motor horn, can be very simply and exhaustively analysed on a type-token basis, because in that case we are dealing with a fixed set of messages which is not subject to contextual variation.

That is one basic distinction between signalling devices. A further possibility that has to be reckoned with is that in the course of evolution certain species may have found it advantageous to develop forms of communication which differ in another way from verbal communication and its analogues: in that they rely not on monomedial signalling but on bimedial or polymedial signalling. That is to say, the signalling units are not, say, acoustic units or visual units or tactile units, but combinations of features from more than one modality. And again, it is possible to envisage mixed or intermediate systems, in which monomedial signals coexist with bimedial signals, for instance.

A third question at this level of generality is whether the signalling is based on semantic structures or semantic isolates. Many of the non-verbal signals used in human communication are based on semantic isolates – waving a hand, or a handkerchief or a newspaper to attract someone's attention for example, is perfectly meaningful, but it is not integrated in any structured way with comparable signals. Whereas human vocabulary, on the other hand, typically tends to be organized around sets of regular semantic contrasts.

These various possibilities are suggested simply by reflection on the fact that even human verbal communication is amply supplemented by, and complexly integrated with, communicational devices which are on the one hand non-discrete and on the other hand polymedial and in some cases not semantically structured. In fact, a great deal of human non-verbal communication – insofar as it has been studied – shows precisely these characteristics.

Reminding ourselves of these possibilities should at least serve as a warning against assuming too readily that what we are looking for in primates is the existence of semantic repertoires based on primitive signals of the verbal type. There seems in fact to be no *a priori* reason why primate communication should have developed in that particular direction.

But the problem is in practice compounded by another factor. So far as we know, human ingenuity has only ever devised two methods of semantic description. One is what we might call the *method of external comparisons* and the other is what we might call the *method of internal contrasts.* In brief, the method of external comparisons is typically the technique adopted by the compiler of a bilingual dictionary who, in effect, compares a supposedly unknown system of meanings with a supposedly known one. Whereas the *method of internal contrasts* is typically the technique adopted by the compiler of a monolingual dictionary, which uses the vocabulary to describe itself and plays off one component against another within one and the same system. The difference between the two methods is like the difference between saying 'A pound is worth 4 DM' and saying 'A pound is worth one hundred pence'. It is very important to keep in mind that those statements are *not* synonymous, even though they may both be true, useful etc. statements about the value of a pound; nor would they become synonymous even if a magical Chancellor of the Exchequer could guarantee that a pound would always be equivalent to 4 DM, as well as being divisible into 100 pence.

Now the method adopted initially by the human investigator of animal semantics has to be the method of external comparisons. The reason why is the same reason as that which imposes this method initially upon missionaries and linguists dealing with an unfamiliar language when translators or interpreters are not available. There simply is no other basis for establishing the preliminary hypotheses of equivalence that are necessary for the process of 'code-breaking'. The drawback about the method of external comparisons, however, is that inevitably it means looking for signals which can be matched up in a convincing way with signals or types of signals from the investigator's own language. This may not matter too much in the missionary situation, since it is a reasonable assumption that the unknown tribe speaks a human language of some kind: that is to say, roughly, a language into which it will eventually be possible to translate the Bible. But where other species are concerned that is not a reasonable assumption. So the investigator appears to be in a predicament. Although he knows it is not reasonable to assume that he is dealing with a communication system structured like a human language, in order to get his semantic description of that sytem off the ground at all he is obliged to look for meanings of the kind that will translate, however crudely, into human terms.

Sometimes in desperation or sheer frustration the investigator tries to take the wrong way out of this predicament. He is tempted, for example, to invoke a simple-minded distinction between 'observer meanings' and 'monkey meanings', as if this disposed of the problem. On the contrary, it makes it worse.

In reports on primate signalling, it is clearly the human observer who is immediately responsible for the assignment of a particular meaning to a particular signal. For instance, if a certain call is described as an 'alarm call', it is being classified semantically in a way which the reporter has decided on, not the monkey. ('Alarm', after all, is a word drawn from the vocabulary of English: it is not a term the monkey uses).

Although no one in practice contests this fact, not everyone is willing to accept the consequences. There are those who want to say something like this: 'When I describe a signal as an "alarm call", although I am responsible for the use of the term "alarm", what I am in fact describing is what the signal means for the monkey.' But when we think through the implications of this claim, it has some very puzzling features. What the claimant is saying is that his report is a report of a meaning-assignment made by the monkey, and not by the reporter. However, it does not seem possible to construe that literally as claiming that the monkey assigned the meaning 'alarm' to the signal, since the monkey does not speak English. So one is led to ask 'What else could the claim mean?'

There are not many possibilities.

It could mean something like: 'If the monkey *did* speak English, he would agree with the meaning-assignment "alarm".' But, short of teaching the monkey English, it is difficult to corroborate such a claim.

Or it could mean something like: 'Although the monkey doesn't speak English, he assigns a meaning to the signal such that if speakers of English assigned the same meaning to it they would assign it the meaning "alarm".' But it is difficult to corroborate this claim either, because it requires that there should be some objective way of comparing a monkey's meaning-assignment and a human being's meaning-assigment, and judging whether or not in a particular case they are the same.

Or it could mean something like: 'The monkey assigns a meaning to the signal such that, to judge by his consequent behaviour, he acts in a way comparable to the way human beings might act if they had received a signal to which they assigned the meaning "alarm".'

But the trouble with this third interpretation is that it is too weak to bear out the reporter's original claim. That is to say, it now emerges that the term 'alarm' does not directly characterize the meaning the monkey assigns to the signal, but simply a comparison between monkey behaviour and hypothetical human behaviour. Again, however, it is not the monkey who is drawing this comparison, but the reporter.

There is nothing wrong with this, of course. But the way to make better descriptions of what monkey signals mean is not to obfuscate the process by attributing your assignments to the monkey. The way to improve a descriptive semantics based on the method of external comparisons is to recognize explicitly the comparative basis of the description, and to try to make the comparisons sharper. That, after all, is precisely what linguists have to do in various forms of comparative linguistics.

Ideally, however, one might hope to be able to take the semantic description a stage further than external comparison with human behaviour allows. That is, one might hope for some progress towards analysing the internal semantic structure of the signalling system. Very possibly, primate signals have not yet been studied in enough detail to bring to light evidence of semantic structuring. But one or two facts are beginning to emerge which suggest that, embryonically at least, primate communication can develop semantic structures.

All good linguists are brought up to believe that the connection between a linguistic sign and its meaning is arbitrary. But they allow that semantic change, on the other hand, is not arbitrary – or at least not arbitrary to the same extent. So one might hope to find in primate signalling evidence of some non-arbitrary relationships which point to the operation of processes of semantic change. I should like to draw attention to just one example which could *prima facie* be interpreted in this way.

In Cheney's paper on vervet monkeys (this volume), we find the following account of a set of acoustically related calls:

> the vervet monkeys' alarm call to a python or poisonous snake is a rasping call (hereafter called a 'chutter') consisting of a series of short, relatively widely separated acoustic units. The energy in these units is concentrated across a wide frequency band, which usually ranges from between 0 to 16 kHz . . . Calls with these acoustic properties are not, however, restricted solely to snake alarm calls. Similar-sounding chutters are used in three other, very different, contexts: when threatening a member of one's own group, when threatening a member of another group, and when threatening the human observer. While chutters uttered in each of these four contexts sound very similar to the human ear, there are small but consistent acoustic differences across different types of chutter . . . In other words, although the calls share similar general acoustic properties, there are nevertheless consistent, significant differences . . .

Now it can hardly be coincidence that the vervet monkey's snake alarm takes the form acoustically of what Cheney describes as 'a rasping call'. This suggests an onomatopoeic origin, of which we find very many examples in the vocabularies of human languages. But nor is it likely to be a coincidence that calls with similar acoustic properties, but certain small acoustic variations, are used as threats: specifically, to threaten members of one's own group, to threaten members of another group, and to threaten human beings. This relationship seems to point to a process of semantic differentiation with a readily comprehensible rationale. What was originally an iconic warning signal is used deliberately to frighten others. However, abuse of this technique, unless checked, simply devalues an otherwise useful signal, as the boy who cried 'Wolf!' discovered too late. So it becomes communicationally advantageous to distinguish acoustically between the warning and the threat in such a way that the warning retains its semantic value, but the threat nonetheless reminds the hearer of an unpleasant association with snakes.

So we can hypothesize a primary semantic differentiation in the history of these signals, which might be represented as follows:

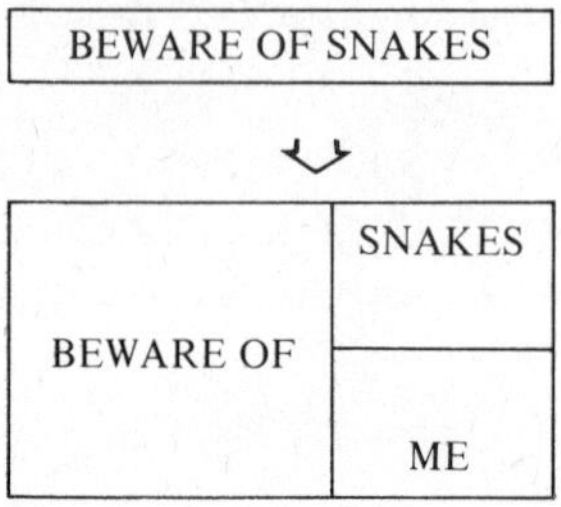

The next stage it makes sense to postulate is a further differentiation of the derived signal, depending on the status of the intended addressee to whom the threat is directed. So a further set of semantic features gets incorporated into the system as follows:

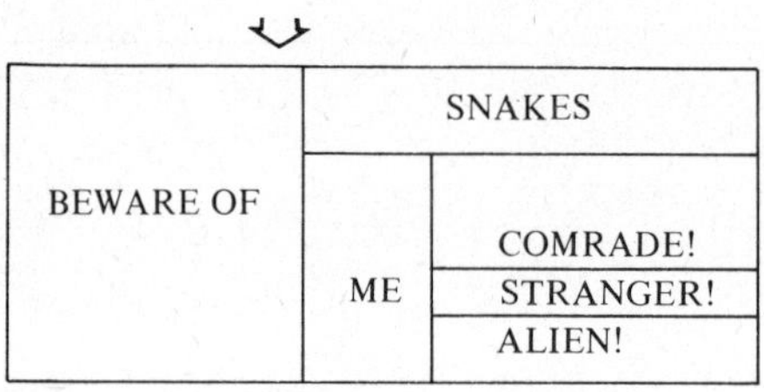

This is a pattern which is very interesting in a number of respects. Analysed in terms of Bühler's three basic semiotic functions (representative, expressive and appellative), it suggests a progression whereby the signal becomes divorced from its original representative function (indicating the presence of a snake), while retaining the original expressive and appellative functions. In other words, the expression of hostility is redirected from referent to addressee, while the addressee in consequence identifies a new source of danger. This in turn establishes a new representative function. It is at this point that acoustic variations in the signal become potentially utilizable to differentiate two representative functions. In this way, the utility of the original signal is not lost, but the communicational repertoire is enlarged.

That developmental pattern in turn is highly suggestive of the ways in which communicational necessities in the lives of animal communities may exert pressure towards distinctive differentiation of signals. It would be interesting to know whether there are other examples which indicate a developmental order 'warning'⟶'threat'⟶'addressee-differentiated threat' and, if so, to study the conditions which appear to favour the emergence of this type of semantic structuring.

It may be that along lines such as these future progress in the study of primate signals lies.

References

Fromkin, V. & Rodman, R. (1978) *An Introduction to Language*, 2nd edn., p. 42. New York: Holt, Rheinhardt.
Harris, R. (1980). *The Language-Makers*, pp. 25–8. London: Duckworth.
Harris, R. (1981). *The Language Myth*. London: Duckworth.
Lyons, J. (1968). *Introduction to Theoretical Linguistics*. Cambridge University Press.

COMMENT

D. PLOOG

Harris seems to make the assumption that non-human primate communication can only be investigated by means of linguistic methods. I doubt that this is true. Nevertheless, I am interested in the two methods of semantic descriptions he has listed. The first was the method of external comparisons where an unknown language is compared with a known language. The second was the method of internal contrast where components of the language are compared with other components of the same language. I am afraid that neither of these methods is applicable in the analysis of primate communication, because there is no other language for comparison, nor are there components to be compared with. We are rather in the position of paleolinguists who want to crack the code of, say, Greek Linear B. We have scribbles, i.e. certain observable behavior units which we call signals (such as a call), which may or may not elicit certain responses of the recipient of a signal. The function of a signal can only be defined by the response of the recipient. At this point of the analysis of the communication system a semantic description would neither be possible nor appropriate.

REPLY

ROY HARRIS

Ploog's distinction between linguistic and non-linguistic methods of cryptanalysis seems to me unacceptable, and is in any case undermined by the parallel he himself draws with Linear B. You cannot analyse a communication system *at all*, whether it be linguistic or non-linguistic, without formulating semantic hypotheses about the signals, however vague or under-determined by the evidence such hypotheses may be. Formal and semantic analysis have to proceed *pari passu*.

8. The inevitability and utility of anthropomorphism in description of primate behaviour

PAMELA J. ASQUITH

8.1. Introduction

The question '*What* do primate signals mean?' sometimes implies the question '*How much* do they mean?'. That is, is the behaviour intentional or purposeful? Anthropomorphism is a concept that many animal behaviourists consider, but mention only in passing when discussing terminology or methodology. Its use has occasionally been defended, but more often warned against by ethologists (for example, Tinbergen 1951; Marler & Hamilton 1966; Morris 1967; Reynolds 1967). Yet anthropomorphism, or the ascription of human mental experiences to animals, occurs throughout professional reports of animal behaviour. It arises through qualitative or ordinary language description in which terms such as 'threaten', 'appease', 'chase', 'greet', 'submission', and so on, are applied to animals' behaviour. These terms already carry meanings associated with human action (that is, purposeful behaviour). The stipulated definitions of these terms by behaviour units which are relatively semantically neutral, such as 'bare teeth' or 'raise tail', do not entirely avoid the purposeful connotations of the sentence 'Animal A threatened animal B'.

Anthropomorphism can be said to occur only if it is presupposed that self-awareness and purposeful behaviour are uniquely human. Cognitive ethologists (Griffin 1976, 1978) and some primate researchers (for instance, Menzel 1975, 1978; Gallup 1977; Mason 1979) have argued against this. However, despite many ingenious experiments that have been devised to test for it, the existence of subjective awareness in animals remains more an ontological than epistemological question. That 'the problem of anthropomorphism' has been of such concern in modern Western science can be traced to the pervasive idea that there is a vast qualitative difference between human and animal mentality. This is in part based on the Christian tradition which emphasized the separateness of humans and animals and on the inheritance of Descartes' (1637) mechanistic philosophy and insistence that only humans possess souls.[1]

Although there is a substantial literature on the justification or otherwise of attributing mental experiences (whether rational or emotional) like those of humans to animals, there has been little attention paid to how that attribution occurs through the nature of language use. Purton (1970) noted that the purposeful connotations of ordinary language terms are retained when used to describe animal behaviour and that this may give rise to anthropomorphism. However, hitherto a mechanism by which this occurs has not been offered. In what follows, I suggest that the mechanism is 'metaphor' which allows and in fact necessitates the semantic link between human action and animal behaviour. It is not my present purpose to discuss whether or not we *should* ascribe purposefulness to animals, but to show firstly that due to metaphor anthropomorphism is unavoidable or inevitable in primate behaviour description, and secondly how anthropomorphism is useful in primate studies.

The following points are presented. A distinction is made between two kinds of anthropomorphism, i.e. generic and specific anthropomorphism. The notion of ordinary language description is explained and Purton's (1970) distinction between A- and O-purposive behaviour is outlined. Four levels of terminology are distinguished, and behaviour units and behaviour categories are discussed in detail. Anthropomorphism is identified as occurring at the level of behaviour categories. The notions of 'behaviour' (movement) and 'action' (meaningful or purposeful movement) are distinguished with a view to characterizing anthropomorphism as the description of animal action. The concept of metaphor is discussed and it is shown how metaphor introduces generic anthropomorphism to behaviour descriptions. The way in which anthropomorphism is a useful aid to theory construction in primatology is explained. Finally, it is pointed out that anthropomorphism is not avoided through use of quantitative description as the latter does not ever completely replace qualitative description.

8.2. General and specific anthropomorphism

Of the many slightly differently worded definitions of anthropomorphism in the literature, that of Katz (1937:19) seems the most unambiguous for the present purpose: to endow an animal with psychical capabilities like those of man and to say that it acts from similar motives. Such anthropomorphism is readily identifiable in descriptions like that by the naturalist McCann, who made the following notes on what he called the common Indian langur (probably *Presbytis entellus*):

> They even made bold to enter the carriages to receive offerings of food and on one occasion I had two great fellows . . . complacently eating the fruit I gave them. They accepted all they got as a right, but resented and were suspicious of any familiarity. (McCann 1928, 192)

An inheritance from Cartesian philosophy is the supposition that humans alone among animals have reasoning minds. Thus, any ascription of purposeful behaviour to animals is anthropomorphism. Katz's definition appears clear enough, but in practice finer distinctions are needed. There are, at an intuitive level, different degrees or categories of anthropomorphic terminology. A distinction is suggested here between general or generic anthropomorphism and specific anthropomorphism.

Specific anthropomorphism refers to the ascription of particular or specific mental and emotional states like those of humans to animals. It is characterized by such description as 'a sentimental elephant' or that of Hornaday who wrote of a captive orang-utan (*Pongo*):

> Dohong was of a reflective turn of mind, and never was entirely willing to learn the things that his keepers sought to teach him. To him, dining at a table was tiresomely dull, and the donning of fashionable clothing was a frivolous pastime. On the other hand, the interior of his cage, and his gymnastic appliances of ropes, trapeze and horizontal bars, all interested him greatly.
> (Hornaday 1922:77)

Specific anthropomorphism is characteristic of some of the earlier Japanese primate behaviour descriptions, for example:

> One of them in particular – a monkey on Takasakiyama called Jupiter – will always be remembered by all of us. He was a monkey of great spirit and valor. Always on the alert, he was strict at times to the point of cruelty. (Itani 1961:424)

By contrast, generic anthropomorphism refers to the ascription of (a general) purposefulness to higher animals, i.e. animals are aware of what they are doing, but it does not imply that psychological experiences are necessarily exactly the same in animals as in man. It includes expressions such as 'threaten', 'appease', as also 'indifferent', 'protect', and so on. In these cases it seems we speak as fellow members of a common communicative system (that we naturally wear King Solomon's Ring). The interpretation of, for instance, a threat by a gorilla is not merely intellectual but intuitive and untutored as well. To some extent, this has a basis in adaptability since threatening signals would be useless as interspecific communication unless recognizable as threatening. These expressions are employed because the animals' behaviour 'reminds us of the comparable human activity, or appears to serve a very similar function in communication. Because they are usually used to refer to human actions they have an inherent purposefulness and it is the ascription of that purposefulness that gives rise to generic anthropomorphism. This more conservative form of anthropomorphism has always been apparent in many scientific studies of primate behaviour. It occurs through the use of ordinary language terms as basic units or elements of the animals' behavioural repertoire and as names for types of social organization.

8.3. Ordinary language terminology

Primatology in America and Europe arose from comparative psychology and ethology (Jaynes 1969; Thorpe 1979). Two inheritances from ethology are study of the development of behaviour in terms of evolution and the importance of field study or the study of the animal's *whole* way of life. These have proved very important in making those researchers aware of the complexities of non-human primate social life. As more and more data have been gathered, enabling more sophisticated questions to be asked about the dynamics of primate social life, descriptions of behaviour have correspondingly employed a wider range of ordinary language terms. It is these terms that give rise to the implication of purposeful behaviour, or a generic anthropomorphism.

By 'ordinary language terminology' is meant terms that are used in ordinary discourse, that are not peculiar to ethology, nor technical neologisms, and that are usually used to refer to human activities. Ordinary language is distinguished from technical language in that the latter is a part of a language (such as English or French) defined only by reference to some particular discipline, occupation or activity among practitioners of which it is current. Technical language consists largely, if not entirely, of vocabulary items that are not part of the related ordinary language; for instance, neologisms such as 'meson' or 'apoblastosis' (Carpenter 1942) or terms not having the same sense in ordinary language as the sense given in a particular discipline (for instance, 'charm' in physics).

Ordinary language in primatology includes the use of such words as 'threat', 'greeting', 'groom', 'hide', 'dominance', 'troop' and 'family'. Some are nouns, some are verbs; some refer to activities and relationships among the animals, others to the type of social organization (which is in effect a summary term for various relationships among the members of a group). The terms are of different levels of descriptive complexity; for instance, 'dominance' is inclusive of more aspects of behaviour and requires more interpretation of sequences of behaviour than does the term 'chase'. There are many ordinary language terms, such as 'nobility', that are rarely, if ever, used to describe animals in ethology, but the selection of appropriate terms is based on little more than a vague apprehension of 'anthropomorphic licence'. Thus, a term such as 'friendship' might not be used, whereas a 'bond'[2] is more acceptable in that it need not imply such a complex arrangement of emotions as does 'friendship'.

However more or less anthropomorphic ordinary language terms are considered to be, they nevertheless reflect assumptions about behaviour based upon experience of human actions. Thus they carry in their meanings something about underlying mental states in humans which is carried over as attributes of the animals. F. V. Smith (1971), a psychologist, has called this 'indirect anthropomorphism', noting that human behaviour is commonly assumed to be purposive and derived from a conscious awareness of the environment and thus that to use the same language to describe the behaviour of animals is questionable for the reasons already stated. He includes as 'questionable' the terms: 'chase', 'intercept', 'elude', 'avoid', and 'pursue' as well as the more explicitly purposeful words 'endeavour' and 'strive'.

The difficulty with ordinary language is that the terms have vague meanings when employed in ethology because they cannot be completely defined in terms of neurochemical control (which is mostly unknown) or movement. The detailed morphology of expressed behaviour differs even among individuals of the same species, thus giving rise to the additional problem of noncomparability of terms based on this sort of definition. This problem was also recognized by human ethologists where, for instance, Grant (1969) noted that if description of behaviour is most acceptable at the level of a complete record of the pattern of muscular contraction, then we would have to recognize, for instance, 'fleeing' by walking rather than by flying or swimming as separate elements. Ethologists have developed more sophisticated techniques to posit some uniformity in the concepts of fleeing, threatening, and so on. They have used for example, causal or functional criteria which can be applied with consistency across species. Thus, 'warning' conspecifics about a potential predator may be recognized (defined) by the function that the behaviour fulfils regardless of how it was manifested (for instance, by shaking branches, calling, running or freezing). However, the problem remains that the existence or nature of the animal's intention to perform the function, or maintain group relations, and so on, though implied by the terms, is not at all clear, or at least is not scientifically testable.

Anthropomorphism arising through ordinary language terminology appears in reports carried out with scrupulous attention to an objective methodology. It occurs simply as a result of the nature of our language, or more specifically, meaning in language. Remarking on the attempt by behaviourists to describe only the outer manifestations of behaviour and to deem inner experience indescribable and thus omit the behaving agent, Midgley noted that this cannot succeed:

> It fails inevitably because most of the terms in which we can describe behavior effectively do refer to the experiences of the agents as well. Reference to a conscious subject always slips in, whatever the disinfecting precautions, simply because language has been so framed to carry it. (Midgley 1978:106)

8.4. Purton's A- and O-purposive terminology

To illustrate how anthropomorphism arises in ordinary language description in ethology, Purton (1970) distinguished between two kinds of purposive terminology. One was 'A-purposive' (A for agent) in which an aware agent does what appears to him to be necessary to attain his goal, and the other was 'O-purposive' (O for organic) which refers to functional explanation of behaviour so that we can speak of the 'goal' of survival without making any reference to purposiveness in the psychological sense. For example, 'aggression' defined as 'striking in order to inflict damage' is very different from 'aggression' defined as 'striking which has the biological function of causing damage'. In the former there is an angry agent, whereas in the latter there is an organism with fixed action patterns of behaviour of which one is classified as 'aggression' since its biological function is to cause damage. When the A-purposive sense of terms is applied to animal behaviour, the terms are anthropomorphic.

Purton outlined how the A-purposive sense arises in biological description:

> Just by looking at the *movements* we cannot tell whether a kick was aggressive or accidental, and the matter is only finally decided when we know whether the *purpose* was aggressive. What is important here is not the movements, or the effect of the movements, but what

> is being *done* in a sense of 'done' which implies that a purposive agent performed an action.
>
> Since we know the sorts of things which animals want and do we inevitably carry over this selection of significant movements into ethology . . . We come to the study of ethology with a background schema of what is important. Yet this schema is a purposive one, and quite foreign to biology or the physical sciences.
>
> (Purton 1970: 40–1)

However, there is a structural analogy between the psychological conceptual scheme and the biological scheme, which can lead to confusions, but which can also be of heuristic value. Purton noted that the successful use of 'anthropomorphic' terminology in biological ethology depends largely on the existence of this analogy. Its heuristic value lies in the familiarity we have with the connections between, for instance, 'attack', 'threat' and 'appeasement', which to a large extent forms the starting point for the ethologist's investigations. Biologists could provide neologisms to describe what they now call 'threat' and so on, but the description would be difficult to understand. Thus, Purton concluded, the value of using purposive terminology is that we can give precise biological descriptions of behaviour (if the terms are understood in their O-purposive sense) which at the same time can be easily 'understood' in terms of ordinary experience. This is because the structure of ordinary language connections remains the same when transferred to the biological framework. This point will become important in the discussion of metaphor.

As noted earlier, anthropomorphism is, in general, eschewed by scientists describing naturalistic primate behaviour. Yet it regularly appears in these reports as what I am calling generic anthropomorphism. This occurs at least partly by virtue of the nature of our language use. It can be shown to arise at a particular stage in the process of describing animal behaviour. To explain, there are different 'levels' of terminology common to primate behaviour reports, of which two – behaviour units and behaviour categories – form the starting point of the following analysis. Behaviour units are basically descriptions of movement patterns and vocalizations (such as 'stare', 'eyebrow flash', 'screech') and behaviour categories are a more inclusive class of terms (such as 'threat', 'courtship', 'appeasement') that are defined in terms of the behaviour units that comprise them. The behaviour units are grouped into behaviour categories according to various criteria (sequential, morphological, functional, etc.: Reynolds 1976), but the movement patterns and vocalizations are the terms of the definition. It is pointed out below that in the move from the descriptive level of behaviour units to the level of behaviour categories, an increment in meaning occurs over and above the meanings of the terms of the definition. Behaviour categories are not simply a shorthand for various movements or vocalizations grouped in one manner or another; they *mean* something more. That is, they imply more than can be ascertained from the movement patterns and vocalizations alone.

The nature of the expanded meaning is a generic anthropomorphism; that is, many of the behaviour category terms imply agency in the animal's behaviour. This occurs because the terms already have established semantic fields in ordinary human discourse in which we normally do wish to impute intentions and feelings to the performer. It is my purpose in the remainder of this chapter to suggest the process by which this anthropomorphic increment occurs.

8.5. Levels of terminology

An ethogram or list of behaviour units with their descriptions is generally given at some point in primate behaviour reports whether in a specific table, in the course of the report, or by reference to previous publications on the species being described. A paper may be published solely for the purpose of giving an ethogram (e.g. Hopf *et al.* 1974), especially if several authors have made separate studies of a species. Many primate ethologists feel there is a semantic neutrality about behaviour units that should be carried over to the more inclusive behaviour categories which are defined in terms of the units. In practice, however, this is not the case.

A distinction is commonly made between the level of behaviour units and the level of behaviour categories. A further two levels of terminology can also be identified. The most basic level, 'below' that of behaviour units, is description of the visible morphology of movement, spectrographs of vocalizations and, in principle but as yet rarely in practice, identification of physiological bases of behaviour. This level is thought to provide the most 'neutral' record of what takes place. It is separate from the next level of behaviour units in that its various components make up individual behaviour units (e.g. Altmann 1962; Strum 1978).

The second level comprises the 'units' of a species' social repertoire which can be understood as the components of the classical ethogram, including such things as facial expressions, postures and named vocalizations. The third level is the more inclusive or general labelling of behavioural categories such as 'dominance', 'courtship', and so on, defined by grouping behaviour units according to various criteria. A fourth level of behaviour that has been suggested is that of social organization in that social organization does not refer to an inanimate object, but is derived from the behaviour, via the relationships, between individual animals (Asquith 1978; Vaitl 1978; Hinde 1979). This is the most inclusive level of behaviour and includes such terms as 'troop', 'harem', and so on. As it is in the move from behaviour units to behaviour categories that anthropomorphism arises, the following discussion focuses specifically on those levels.

A behaviour unit, variously called behaviour item, behaviour type, taxonomic element (Reynolds 1976), or behaviour pattern (Altmann 1962), has been defined as an essential or core movement (*Kernbewegung*) which can be accompanied by typical but indispensable movements (Kummer 1957). Wiepkema (1961) selected behaviour units to be (*a*) easily measurable, (*b*) not too rare in occurrence, (*c*) biologically meaningful and (*d*) not entirely correlated with other variables. Slater (1978) felt that a behaviour unit should be (*a*) species typical, (*b*) made up of movements that occur simultaneously and sequentially with a high degree of predictability and (*c*) repeatedly recognizable.

An example from a table (Strum 1978) that includes the first three levels of description will serve to illustrate various points. Only part of Strum's list will be given. She named her table 'agonistic behaviours' and within that included aggressive and submissive behaviour. All three terms are behaviour categories, but agonistic behaviour is a more inclusive or complex term than the other two. Under aggressive behaviour, Strum listed the behaviour units that occur in or make up aggressive behaviour, along with an accompanying list of behaviour movements which she calls description (of the units). She also distinguished between categories of low, medium and high intensity aggression: Table 8.1 is taken from the low intensity aggression category for *Papio anubis*.

Hinde (1973) pointed out there are two basic methods for describing behaviour units: (*a*) physical description or in terms of patterns of limb or body movements (e.g. 'knee jerk', 'species-characteristic song') or (*b*) by consequence, which covers all patterns that lead (or could lead) to a specified result (e.g. 'picking up nest material', 'approaching'). Description by consequence provides a useful shorthand, but can also be subdivided into physical description. It will be seen in Strum's list that most behaviour units (for example, 'directed stare', 'head bob', 'tail up') are quite neutral in meaning though occasionally an interpretation as to the meaning of the behaviour units creeps in as purposive terminology (such as 'fear face', 'active avoid', 'hide'). This corresponds to the two kinds of description mentioned by Hinde above. At this level, this is

Table 8.1. *Behaviour units of low intensity aggression*

Aggressive behaviour	Description
Low intensity aggression	
Direct stare	Obvious
Raised eyebrows	Sometimes accompanied by lowered eyelids
Ears flattened	Ears pressed against head
Head bob	Jerking head down and forward
Yawn	Mouth opened exposing canines to varying degree depending on intensity – oriented towards another individual
Pant grunt	Loud two-phased grunt towards another individual
Submissive behaviour	
Avoid a stare	Glancing away from another individual or fixation on the ground
Fear face	Grin or grimace
Cough geck	Single sharp sound with cough-like quality often accompanied by jerks or body twitches
Tail up	Tail is raised and held in a vertical position
Active avoid: fearful	Running away with combination of geck, scream, fear face and tail up
Crouch	Body lowered to the ground, orientation to another individual
Hide	Avoid contact by hiding out of sight – adult males frequently use pig-holes
Approach: fearful	Oriented approach with tail up and fear face, geck or scream

After Strum (1978).

usually easy to 'catch' and may be qualified at the time or enclosed in quotation marks to show that the author is aware of the implication. Altmann (1962) treated purposive units this way in his catalogue of rhesian behaviour. Thus, in his behaviour unit 35: Looks 'apprehensively' (toward), Altmann noted:

> While the term 'apprehensive' is anthropocentric [*sic*], it may convey best the impression of an individual who glances quickly around him every few seconds, seems to be tense, to have a low avoidance threshold, and so forth. These responses may or may not be directed toward another individual. (Altmann 1962:378)

Similarly, with behaviour unit 54: Ignores, he stated.

> It was sometimes obvious that a monkey saw, heard or felt a behaviour pattern that was directed conspicuously toward it, yet 'ignored' the other monkey or monkeys involved, i.e., continued doing what it had been doing. (Altmann 1962:379–80)

The neutrality of behaviour units is stressed. Box & Pook stated that:

> It seems to us that at least at the onset of a study of social behaviour, it is methodologically advantageous to be relatively neutral, and to record the behaviour of individual animals, rather than to start with the complex units of social inference. (Box & Pook 1974:104)

The actual naming of behaviour units is not intended to give social meaning to the movement patterns, but rather to serve as a sort of mnemonic device. As Altmann stated:

> The titles and descriptions of the behavior patterns (units) are not intended as definitions, but are presented, rather, in the hope that they are sufficient to give the reader some sort of picture of the constituents of rhesian behavior and to enable others who work with members of this or related species of primates to identify the behavior patterns when they observe them. Those behavior patterns that commonly precede or follow each behavior pattern have not been included always in the descriptions. While the justification for including any particular pattern in the catalogue is to be found in the social context in which the pattern occurs, this social context is never part of the definition of the pattern.
> (Altmann 1962:374)

Likewise, Hopf (1972:366) stated 'The names of the units should be regarded as short and handy labels, not as an *a priori* interpretation'.

Hopf *et al.* (1974:225) felt it was indispensable that behaviour units be defined in the sense of qualifying the terms' usual meanings in ordinary language: 'It often leads to misunderstanding when words of everyday language are used for behavioural terminology without additional information concerning distinguishing criteria'. Marler & Hamilton stressed neutral description of behaviour units:

> In what terms should these action patterns be described? Needless to say they should be objective. Anthropomorphisms should be rigorously excluded. Rather than using such terms as 'afraid' and 'angry' which involve both observation and interpretation, every effort should be made to record the spatial coordination of limbs and body from which the action pattern is constructed. Sound may be recorded or noted onomatopoetically, odors by analogy with other odors. To complete the basic description the temporal organisation of these elements should also be noted.
> (Marler & Hamilton 1966:715)

Not only emotional terms such as 'angry' and 'afraid' are, however, anthropomorphic in this context. As Purton pointed out:

> The more one gets rid of the guide-lines of purposive terminology

> the more one is likely to end up with an unorganised heap of observational and experimental data. On the other hand, if one is selective in choosing what to observe this selection is likely to be strongly influenced by one's awareness of what the animal is *doing*, in a sense of 'doing' which presupposes that the nature of the action is defined by the animal's purpose. It is not enough just to eliminate references to 'fear', 'anger', etcetera, since expressions such as 'jumping over the fence', 'looking up', 'picking up a seed', are equally 'anthropomorphic'. This is obvious when we consider human beings – there is a clear difference between a person jumping over a fence, and his body moving over the fence. The first statement implies the second, but the second does not imply the first: the body may move over the fence because it has been thrown, or because of a violent muscle spasm, perhaps. In such cases the moving over the fence is not something which the person *does*.
> (Purton 1970:152–3)

The difference in meaning implied by a person (or animal) jumping over a fence and a person's (or animal's) body moving over a fence is implied by the terms 'action' and 'behaviour' respectively (as, for instance, in Weber 1947; Harré & Secord 1972). These terms are introduced below.

8.6. Behaviour and action

The term 'action' has been used with a specific meaning in some primate behaviour catalogues. For instance, in a list of methodological definitions at the end of a study, Hopf defined 'action' as:

> Action (behavioral): a behavioral unit displayed by an actor (to a recipient if unit is partner directed) at a certain time; and for a certain duration. Type of action: combination of a behavioural unit, an actor, and a recipient (if interactional), independent of actual occurrence; in a mathematical sense: the variable, while the action represents the value. (Hopf 1972:387)

Pruscha & Maurus defined 'action' as:

> each behavioural event . . . is classified according to 'who does what to whom': a sender animal (*a*) performs towards a recipient animal (*b*) a behaviour unit (*u*). We will call a behavioural event coded in this way a triplet (*a*, *b*, *u*), an action (when emphasizing its temporal uniqueness) or action unit (when emphasizing the type of recurring event). (Pruscha & Maurus 1976:187–8)

In the present chapter, the term 'action' as distinguished from 'behaviour' will be employed differently from the meanings given above.

The utility of the different usage lies in the kind of distinction it provides. 'Behaviour' will refer to movement or physical patterns only and 'action' will refer to purposeful behaviour. Thus 'action' refers to another dimension in addition to the description of movement patterns. This becomes significant in the discussion of the basis for anthropomorphism in behaviour reports.

Reynolds has summarized Weber's (1947) distinction between 'behaviour' (*Verhalten*), 'action' (*Handeln*) and a third term, 'meaning' (*Sinn*), which links the two others:

> If we describe what people or animals do, without inquiring into their subjective reasons for doing it, we are talking about their *behaviour*. If we study the subjective aspects of what they do, the reasons and ideas underlying and guiding it, then we are concerned with the world of *meaning*. If we concern ourselves both with what people are, overtly and objectively, seen to do (or not to do) and their reasons for so doing (or not doing) which relate to the world of meaning and understanding, we then describe *action*.
>
> (Reynolds 1980:xxiii)

Harré & Secord (1972:39)[3] gave examples to illustrate the distinction between behaviour and action (in each case, (*b*) is an example of an action):

1. (*a*) His arm extended straight out through the car window.
1. (*b*) He signalled a left turn.
2. (*a*) Her arm moved rapidly forward and made contact with his face.
2. (*b*) She slapped him angrily.

They made several points about action: action has significance and meaning; it occurs in a social, not a physiological context; it is inextricably bound up with the nature and limits of language (and the fabric of society), and, importantly, there is no way of reducing action to movement and so of setting it within a physiological context. To try to do so is to transform what was action into something else.

This has obvious implications for the definition of behaviour units and categories. If, when appearing in the body of a report, the categories become read or understood as action (which is what has happened when the terms are said to be anthropomorphic), the meaning the terms acquire cannot be translated into or defined in their apparently original form (i.e. of movement patterns). As Purton (1970) remarked, a description of animal 'behaviour' would not make sense.

To summarize, ordinary language terms are sometimes read as referring to animal *action* in primate behaviour reports. A transformation of meaning occurs when behaviour categories are removed from their tables and become part of the flow of description about activities in an animal group. Very often, lists of behaviour units that make up the categories are not included in the reports, and the reader is left to determine these units in the course of the report. Theoretically, one should be able to make a list of the categories and the behaviour units by careful reading or dissection of the text. However, it becomes apparent in the course of such analysis that the whole sense of the original report is not retained when the categories are again reduced to their component units. Further, even where definitions of behaviour units and categories are stipulated by the writer, use of the ordinary purposeful meaning of those terms is often apparent.

A field study provides perhaps the sternest test for the adequacy of the primatologist's definitions because animal behaviour in the wild is more 'rounded off' within a richer context than in a laboratory setting. The display of much more of the animal's social repertoire and effects of its richer social and physical environment give a fuller and at the same time less easily analysable picture of behaviour. Discrete definitions of behaviour categories are less easily applied in the flow of behaviour. Albrecht & Dunnett remarked that the complexities of chimpanzee behaviour as observed in the wild made interpretation of the behaviour difficult:

> The social behaviour of chimpanzees does not lend itself well to division into rigid categories. This particularly is true of sexual behaviour, which appears to be incorporated in a variety of social behaviour patterns apart from straight-forward copulation.
>
> (Albrecht & Dunnett 1971:30)

and that,

> The division of behaviour which appears in the following subsections is therefore somewhat arbitrary. The material has been presented in what seemed to the authors to be the most natural fashion and the reader should therefore not be surprised if he finds in a certain subsection material which does not seem strictly to come under the title of that subsection.
>
> (Albrecht & Dunnett 1971:31)

Behaviour categories are usually defined analytically as being made up of various behaviour units. Once in the field context, however, it becomes possible to employ other methods of definition as behaviour categories exist at different levels of inclusiveness. For instance, 'threat' can be part of 'dominance' and 'dominance' and 'submission' can be elements in the description of 'agonistic behaviour'. In the second quote from Albrecht & Dunnett it is evident that observers must at times rely on intuition (i.e. 'what seemed to the authors'), rather than analysis, to determine to which category some of the behaviour units or less inclusive behaviour categories belong.

An example of this occurs at the end of a report on the behaviour of bonnet macaques (*Macaca radiata*) (Simonds 1965) in which is written:

> The less dominant males were less active generally. And there was a personality difference, as well, which may have been a result of their subordinate position rather than the cause of it. For instance, one of the older and smaller males never initiated a threat sequence without the support of a more dominant male, but he would join any threat sequence in which other males threatened those immediately below him in dominance. After such a sequence he always chewed on the neck of the subdued monkey. He could be characterized most simply as a cowardly bully. (Simonds 1965: 185)

Characterizing a monkey as a cowardly bully (even as a convenient shorthand) on the basis of when and how it performed its threatening sequences implies some meaning for threat other than the physical descriptions given in the body of Simonds' report. Robinson commented on what is actually the semantic field of terms:

> A word has other dimensions besides its indicativeness. Besides indicating it expresses. A word for *x* indicates *x* and expresses the idea of *x*. When a man says 'grey', he indicates grey and expresses his idea of grey. Every word carries with it, in each speaker's mind in which it lives, certain associations other than its main association to what it indicates, associations to other things or associated emotions and attitudes . . . often the same association holds for nearly all speakers of the language. (Robinson 1972:56–7)

The usual associations of ordinary language terms are not destroyed in stipulated definitions. Robinson made a very relevant comment about this:

> In stipulating a meaning for a word, a writer demands that his reader shall understand the word in that sense whenever it occurs in that work. The writer thereby lays upon himself the duty of using the word only in that sense, and tacitly promises to do so, and tacitly prophesies that he will do so. But sometimes a writer does not use the word only in the sense he has stipulated. Then his stipulation implied a false premise and a false prediction . . . We often find a writer evidently using a word in a sense other than the sense he stipulated earlier in the work . . . It is not necessarily a case of dishonesty. Or at any rate the dishonesty may be in the words themselves rather than in the writer; for words often deceive their own utterers about what they mean as they are uttered!
>
> (Robinson 1972:64–5)

It is the purpose of the next sections to explain one characteristic of language, metaphor, that reduces the effectiveness of the primatologists' stipulated definitions of ordinary language terms in avoiding anthropomorphism.

8.7. Metaphor

The use of ordinary language terminology to describe animal behaviour has been characterized as being metaphorical (e.g. Purton 1970; N. Tinbergen personal communication). That is because these terms carry with them associations of conscious agency which are not, in the context of scientific behaviour studies, meant to be applied literally to the animal's behaviour. Thus, anthropomorphism is generated. Purton wrote:

> These words (aggression, threat, courtship, appeasement, begging, and so on) have a decidedly anthropomorphic sound, but it is clear from any but a very superficial reading that most of the biological ethologists are not anthropomorphising. Their language of 'threat', 'appeasement', etcetera, is metaphorical, in the sense that they are not concerned with desire and action at all. The framework within which they work has no place in it for any notions that presuppose consciousness; it is a strictly biological framework.
>
> (Purton 1970:110)

The application of the concept of metaphor in this context, however, is not quite as straightforward as it at first appears. For instance, though there would be no question that the terms were used metaphorically if electric toy animals were so described, there is some sense in which these terms are being used literally of living animals. Besides (and perhaps because of) the fact that many of us mean to ascribe intentions to animals in everyday life, we also use ordinary language terms in scientific discourse about animals because they are *appropriate* for such description.[4] Their suitability stems from the fact that living creatures are being talked about of whose behaviour we feel some intuitive understanding in everyday experience (based, perhaps, on a common biological heritage).

There is moreover an inevitability (or lack of any real alternative) about the overall use of ordinary language terms. Purton (1970) pointed out that these terms are heuristically useful in providing a ready-made framework in which to understand animal behaviour. But this has a price as the user is bound by the structure of the ordinary mental connections that are made between these terms. Due to our ordinary use of terms, a concept such as 'threat' must be associated with certain others such as 'attack', 'appease', 'retreat', 'flee'. Thus, once the ethologist has named a behavioural item a 'threat', his naming of the reaction or preceding behaviour tends to be predetermined from ordinary use. That is, we do not ordinarily associate 'threat' with 'grooming' (unless it is qualified as 'appeasement grooming') or with 'begging', and it is equally nonsensical in the scientific case. Nor would the description read sensibly if a realm of discourse other than ordinary language were suddenly employed in mid-sentence. In true metaphor, however, this sort of restriction on choice of words does not apply.

Nonetheless these qualifications by no means necessitate the abandonment of the metaphor concept as giving rise to anthropomorphism in ethology. They merely point to the fact that something may be occurring in addition to making a metaphor in ordinary language description of animal behaviour. Nor is it the case that ordinary language terms create only a simile, because by saying 'one dog threatened another dog', we are not in our minds simply listing similarities and differences between human threat and animal threat as would be the case in simile. Rather, there is an additional sense or meaning given to the description that would be absent in another possible kind of description, such as one in terms of movements. Thus, the description, 'one dog bared its teeth, pricked its ears and made a guttural noise in its throat in the presence of another dog' means something different from 'one dog threatened another dog'. The nature of the 'additional sense' and how it arises is the subject matter of metaphor.

It will be useful to describe a few notions of metaphor in some detail, both to formulate more explicitly the 'additional sense' that it generates and thus explain how anthropomorphism arises unsolicited in the special case of ethology, and to explain other important ways in which metaphor 'works' in primate ethology. The importance of metaphor as a powerful tool of research in generating theories about 'second-order' behaviour (i.e. social structure) will be suggested.

8.8. Concepts of metaphor

Those who have analysed metaphor have posed various and overlapping questions about it, such as McCloskey's (1964:215) 'What do we do when we use words metaphorically?' and 'How does a metaphorical word function?', and the several questions posed by Black (1962:25), among them: 'How do we recognize a case of metaphor?', 'In what sense, if any, is a metaphor "creative"?', or 'What do we *mean* by metaphor?'. It seems, however, that most of the important points about metaphor can be considered by addressing a paraphrase of McCloskey's questions: that is, 'What *happens* when we use a metaphor?' A distinction is generally made between the literal and metaphorical use of words. As Black assumes an intuitive grasp of what the literal use or meaning of a word is, it is worth mentioning one or two distinctions drawn by McCloskey because the literal and metaphorical meanings of words are basic to much of the subsequent analysis of metaphor. The literal meaning of a word, according to McCloskey, is that which is agreed upon or conventional in a group with a common language. A literal meaning can only be right or wrong with reference to proper, that is, common or accepted usage. On the other hand, a word used metaphorically cannot be corrected by reference to proper usage, but can only be criticized as inappropriate or inept. There is no agreed upon or definitive rule for using a word metaphorically, but it will only be appropriate if the meaning of the 'metaphorical word' can in some sense be associated with the meanings of the other (literal) words in the sentence. That 'association' of meanings will be discussed more fully presently.

Another distinction drawn by McCloskey is that of the parasitic relationship between the metaphorical use and literal use of words; that is, the metaphorical use presupposes the literal one, but not vice versa. In other words the force of metaphor is derived partly from the literal meaning of the word used metaphorically but no literal meaning is derived from a metaphorical one (except in the special case of catachresis), wherein a metaphor provides a meaning or sense that literal words do not convey. As defined by Black (1962:33), '"catachresis" [is] . . . the use of a word in some new sense in order to remedy a gap in the vocabulary; catachresis is the putting of new senses into old words. But if a catachresis serves a genuine need the new sense introduced will quickly become part of the literal sense. Thus, in the metaphor 'cotton-wool clouds' McCloskey explained:

> Looking white and fluffy are striking characteristics of cotton-wool, which, when applying the literal word 'cotton-wool', we take as signs for all the rest of the characteristics of cotton-wool; but when we apply the word 'cotton-wool' metaphorically to clouds, we are no longer using them as signs. The things called by the literal word have the striking characteristics together with a lot more; whereas the things called by the metaphorical word, have only the striking characteristics.
>
> (McCloskey 1964:229–30)

Thus it is the striking but superficial characteristic that the metaphorical use of a word calls to mind. How does this happen? 'A metaphor is a word for one sort of thing applied to a different sort of thing, so a metaphorical word is applied differently from a literal word: it is applied in virtue of different characteristics' (McCloskey 1964:227). McCloskey distinguished between two ways in which the metaphorical use of a word relates to its literal use: (*a*) the metaphorical use may be logically related to the literal use in being partially, but not wholly, identical with it, or (*b*) the metaphorical use may be logically distinct and only psychologically or contingently related to the literal use. This distinction is made in terms of the 'paradigm characteristics' of a word. The paradigm characteristics are those which justify the literal application of a word (what makes one cloud like another cloud). Some metaphorical words share in some of the paradigm characteristics of the word; others do not when the metaphorical meaning is entirely different from the literal one.

McCloskey gave an example of shared paradigm characteristics in the metaphor: 'A bright low ribbon of rainbow'. Here, the metaphorical use of 'ribbon' is similar to the paradigm characteristics of ribbons in being relatively narrow – a band – and being brightly coloured. An example in which no paradigm characteristics are shared is given by: 'The pond, I suppose from over-pressure when it was less firm, was mapped with a puzzle of slight clefts branched with little sprigs' (McCloskey 1964:227). In this example, the metaphorical use of maps and sprigs recalls none of the usual characteristics by which we identify them as maps and sprigs. As McCloskey pointed out:

> Being made up of a mass of interlaced lines, or being of a certain shape, are not among the paradigm characteristics of maps or sprigs. Yet they are the characteristics in virtue of which the appearance of the pond resembles maps and sprigs, and in virtue of which the metaphorical use is made possible.
>
> (McCloskey 1964:227)

Thus, the metaphorical is distinguished from the literal use of words in that the former is neither correct nor incorrect with reference to common usage; the metaphorical is parasitic (except in catachresis) upon the literal use of words, and the metaphorical shares in some or none of the characteristics of the thing to which a word refers – characteristics that make the use of that word necessary in literal parlance.

A metaphorical word has a certain semantic relation to the literal meaning of that word and to the other literal words in a sentence. That association of meanings, or, more precisely, between those things the association is taken to hold, forms the basis of the three views of metaphor discussed by Black (1962). The quality and direction of the association goes some way towards answering the question of what happens when we use a metaphor.

In his analysis, Black distinguished between the 'frame' and the 'focus' of a metaphor. In the metaphor 'The chairman plowed [*sic*] through the discussion', our attention is drawn to or focuses on one word, 'plowed'. In this example, 'plowed' is the focus of the metaphor and the rest of the sentence comprises the frame. (In McCloskey's usage, the frame is the context of the metaphorical word.) In calling a sentence a metaphor, Black (1962:27–8) wrote, we are implying that at least one word (here, the word 'plowed') is being used metaphorically in the sentence, and that at least one of the remaining words is being used literally. Given a particular focus, only certain frames produce a metaphor. In other words, the presence of one frame can result in metaphorical use of a word, while the presence of a different frame for the same word fails to result in metaphor. To call a sentence an instance of metaphor is to say something about its meaning, not its orthography or grammatical form. Thus, if a sentence that is a metaphor is translated word for word into a foreign language for which this is possible, we would say the translated sentence is the same metaphor. Any part of speech can be used metaphorically (nouns, verbs, adjectives, etc.). Meaning, rather than grammar, is the relevant factor in metaphor.

Black presented three views of metaphor, based on the interplay between three meanings present in any metaphor: (*a*) the metaphorical meaning of a word, (*b*) the literal meaning of a word that is being used metaphorically and (*c*) the meaning of other literal words in the metaphor (the frame). The first is the 'substitution view' of metaphor. This view holds that a metaphorical expression is simply a substitution for an equivalent literal expression. So that in the metaphor 'Richard is a lion', 'a lion' simply means 'brave'. The reason given on this view for the use of metaphor at all is either for catachresis or to give pleasure to the reader in having to solve a riddle or having a diversion to a concrete image (in this instance, a lion).

The 'comparison view' of metaphor maintains that metaphor consists in the presentation of the underlying analogy or similarity. In this view, 'Richard is a lion' is approximately equivalent to 'Richard is like a lion'. The comparison view is a special case of the substitution view. As in the substitution view, the metaphorical statement is taken to be standing in place of some literal equivalent, but the original statement is treated as a more elaborate paraphrase in that it is interpreted as being about, for instance, lions as well as about Richard.

A third view is the 'interaction view' of metaphor, which is the one

Black favours as offering insight into the uses of metaphor. He stated: 'when we use a metaphor we have two thoughts of different things active together and supported by a single word or phrase, whose meaning is a resultant of their interaction' (Black 1962:38). In the metaphor 'The poor are the negroes of Europe' an interaction view holds that our thoughts about European poor and American negroes are active together and interact to produce a meaning that is a resultant of that interaction. Thus, the focus 'negroes' obtains a new meaning which is not quite its meaning in literal use, nor quite the meaning any literal substitute would have. The context or frame imposes extension of meaning upon the focal word. The reader must attend to both the old and new meanings together. Further, the literal words in the frame have their meanings altered somewhat. That occurs because the metaphorical word causes us to think of only certain characteristics normally associated with the literal words and so reorganizes our view of the literal meanings. Thus, in the example 'man is a wolf', the commonly held characteristics of wolves will be applied, as far as possible, to the principal subject, man. Black wrote:

> But the implications will *not* be those comprised in the commonplaces *normally* implied by literal uses of 'man'. The new implications must be determined by the pattern of implications associated with literal uses of the word 'wolf'. Any human traits that can without undue strain be talked about in 'wolf-language' will be rendered prominent and any that cannot will be pushed into the background. The wolf-metaphor suppresses some details, emphasizes others – in short, organizes our view of man. (Black 1962:41)

Similarly, 'if to call a man a wolf is to put him in a special light, the metaphor also makes the wolf seem more human than he otherwise would' (Black 1962:44).

Black also pointed out that the use of metaphorical language regularly produces shifts in attitudes. Thus a wolf is (or was) conventionally a hateful and alarming object; so to call a man a wolf is to imply that he too is hateful and alarming. Similarly, if a battle is described purely in terms of a game of chess (in which all expression of feeling is formally excluded), it excludes, by choice of language, all the more emotionally disturbing aspects of warfare.

The substitution view and comparison view of metaphor might suffice for quite trivial instances of metaphor, but they can be replaced by literal translations (except possibly in catachresis) with no loss of cognitive content. However, 'interaction-metaphors' require the reader to use a system of implications. A secondary subject (the metaphorical word) is used to give insight into the principal subject (the literal word(s)). It demands simultaneous awareness of both subjects and is not reducible to any comparison between the two. A set of literal statements about the 'interaction-metaphor' will not have the same power to inform and enlighten as the original. For instance, in a literal paraphrase, the implications – previously left for the reader to work out himself with a feeling for their relative priorities – are now expressed explicitly as though having equal weight. The literal paraphrase inevitably says too much, and with the wrong emphasis. Black stressed that this constitutes a loss in cognitive content – the weakness of the literal paraphrase is that it fails to give the insight that the metaphor did. Wheelwright (1954:97) may have been implying something similar in his observation that 'metaphor is a medium of fuller, riper knowing, not merely a prettification of the already given'.

An interaction view of metaphor is important to the analysis that follows. The way in which Black presented this view, however, has been criticized by those who would reduce metaphor to a kind of simile and argue that metaphor does not add meaning to the subjects of the metaphor. A difficulty with Black's analysis is his principle that in the interaction view each metaphor has two distinct subjects, and it is the two subjects, or their systems of implications, which interact. Thus, his theory is made applicable primarily to those metaphors which involve two nouns, and inevitably suggests comparisons. His particular theory cannot handle such metaphors as 'the eddying brink' in which two subjects are not even present to the mind. Nonetheless the idea that the combination of meaning effected by a metaphor results in a new and unique agent of meaning is accepted by the majority of modern commentators (Martin & Harré 1982).

Although Black's scheme is adequate for the animal behaviour case, this criticism should first be met. The difficulty is overcome by reference to Richards's (1936) discussion in which the interaction view was originally put forward. Richards (1936:93) noted that, 'when we use a metaphor we have two thoughts of different things active together and supported by a single word or phrase, whose meaning is the resultant of their interaction'. Black dismissed the notion of 'two thoughts working together' as psychologistic. Martin & Harré (1982) pointed out, however, that as the thoughts one is dealing with here are primarily couched in words, Richards's terminology is no less convenient than Black's 'two present subjects'. Richards emphasized that metaphor is an interaction of thoughts, not a mere shifting of words. His distinction between 'tenor' and 'vehicle' is based on the notion that two thoughts interact in metaphor. The tenor is the underlying subject of the metaphor and the vehicle is the terms which present it. In the metaphor 'A stubborn and unconquerable flame creeps in his veins and drinks the streams of life', the tenor is the fever from which the man is suffering, and the vehicle is the flame which drinks his life. Martin & Harré pointed out:

> Here it is important to note that what Black would call the primary subject, 'fever', is not explicitly mentioned in the passage. We talk about fever by using the word 'flame' and its associations determine what we mean by so doing. Since 'flame' occupies the centre of a different semantic field from that occupied by 'fever', the use of the term 'flame' enables us to say things about fever, different from those we could say by using the word 'fever'. This supports Richards's contention that it is thoughts (associated commonplaces) and not words, which interact. (Martin & Harré 1982:6)

Thus we are not dependent on two present subjects in a metaphor to get an interaction of meanings between their respective semantic fields. The tenor and vehicle may be co-present in one word or phrase (as in 'That wolf is here again'), since tenor corresponds to 'reference' and vehicle to 'sense'. A full interaction theory is then also possible in a metaphor that has one true subject.

A number of these aspects of metaphor discussed above help illuminate the consequences of the metaphorical use of ordinary language terms to describe non-human primate behaviour.

8.9. The metaphorical basis for anthropomorphism

Examples of metaphor of the kind suggested by Purton (1970) are quite widespread in primate behaviour reports. His distinction between A-purposive behaviour (behaviour performed by a conscious agent) and O-purposive behaviour (nonintentional 'behaviour' of an organism) provided a useful framework with which to describe two realms of discourse or ways to 'read' descriptions of animal behaviour. An O-purposive reading gives a functional account (which can be used to explain the behaviour's usefulness for survival of the individual and species) whereas an A-purposive reading gives a psychological or anthropomorphic account.

An interesting paper by de Waal & van Roosmalen (1979) on 'reconciliation' and 'consolation' among chimpanzees provides a clear example of this. Part of their abstract described the purpose of their study as follows:

> After agonistic interactions among chimpanzees, former opponents often come into non-violent body contact. The present paper gives a

> *quantitative* [my emphasis] description of such contacts among the chimpanzees . . . in order to establish whether these post-conflict contacts are of a specific nature.
>
> (de Waal & van Roosmalen 1979:55)

Their results showed that former opponents tended to contact each other relatively shortly after a conflict and that they showed special behaviour patterns during these first contacts. Contacts between the aggressed chimpanzee and a third party also occurred with greater frequency shortly after a conflict. Behavioural items characteristic of these contacts were 'kiss', 'embrace', 'hold-out-hand' (begging), (de Waal & van Roosmalen 1979:64), 'submissive vocalization' and 'touch'. The relative ratios of these behaviours differed between contact of former opponents and contact with a third party. For instance, the kiss/embrace ratio for inter-opponent contacts was 23/19 and for third party contacts 5/37 (Table 1 in de Waal & van Roosmalen 1979:60). The authors stated:

> As post-conflict contacts involve special behaviour patterns we may give them special names. First inter-opponent contacts will be called 'reconciliation', and contacts between the aggressed party and a third animal shortly after the conflict will be called 'consolation'. For reconciliation, 'kiss' is the most characteristic element, whereas 'embrace' is so for consolation.
>
> (de Waal & van Roosmalen 1979:64–5)

The findings are based upon quantitative data and the authors noted that the terms 'reconciliation' and 'consolation' are objectively definable interaction types, as indeed they are. However, they also added:

> the terminology clearly is functional rather than descriptive. These terms reflect our impression that such body contacts have a calming effect and serve an important socially homeostatic function.
>
> (de Waal & van Roosmalen 1979:65)

Thus, that reconciliation (or consolation) occurred between two chimpanzees can be read as an O-purposive statement, objectively defined and as serving a specific function. However, the A-purposive connotations of those particular words are not only present in the minds of the authors (who enclose the terms in quotation marks) and of the reader, but add a mentalistic interpretation of what is occurring between the animals: i.e. 'These terms reflect our *impression* that such body contacts have a calming effect'. The system of associated meanings in the human experience of reconciliation and consolation are both a reflection of and exist by virtue of the choice of terms.

The metaphor thus reads, 'After a conflict, a reconciliation often occurs between two chimpanzees.' The authors have not included this particular statement in their report, but it would follow from their own choice and definitions of terms. A specific statement they did make (de Waal & van Roosmalen 1979:64), which is another instance of this sort of metaphor, is: 'the data seem to indicate that after an agonistic interaction many participants *prefer* contact with the opponent over contacts with other group members.

For the purpose of explaining, through the concept of metaphor, the process by which A-purposive connotations slip in, a more simply stated metaphor will be useful. Thus, in 'A young chimpanzee greeted its mother', 'greeted' is the focus of the metaphor and the rest of the sentence is its frame (in Black's (1962) terminology). In its ordinary or literal use, 'greeting' connotes a complex set of meanings suitable to the various occasions, manners and purposes of greeting in human experience. In the metaphor above, 'greeted' means something more than, for instance, 'A young chimpanzee touched its mother on the thigh', etcetera. It can be asked if the language for human activities serves a catachretic function (fills a gap in our vocabulary) in description of animal behaviour. If so, we might ask why there is a lack in our vocabulary; why description in terms of movement or function only is not sufficient for understanding primate behaviour. This brings us back to the original point about the appropriateness of ordinary language terms and the fact that no other modes of description make animal behaviour as understandable to us. Further, the fact that the structure of connections we make between ordinary language terms, such as between the concepts of 'threat', 'appease', 'attack' and so on, is necessarily carried over unaltered to the animal case, means that the metaphor is propagated throughout a report.

Certain characteristics of metaphor are illuminating on these points. The metaphorical use of a word presupposes the literal use and what the metaphor calls to mind are the striking characteristics of the literal use of the word (McCloskey 1964). Evidently, one of the striking characteristics about the word 'greeted' and other ordinary language terms is their implication of a conscious agent performing actions, aware of what he is doing. That this is the case is shown by the fact that people say these terms are anthropomorphic when they are applied to animal behaviour.

McCloskey (1964) has pointed out that a metaphorical word will only be appropriate if its meaning can in some sense be associated with the meanings of the other (literal) words in the sentence, and it is the striking characteristics that are so associated. Thus, for the purpose ordinary language terms to be appropriate to their animal context, there must be some underlying feeling that the animals to which the behaviour is attributed are agents. This seems to be what is in fact implied, and the basis for the charge of anthropomorphism in the use of this terminology.

Another characteristic of metaphor is the tension it creates between the common or literal use of a word and the unusual or metaphorical use. McCloskey (1964) noted:

> Because the language is a common language commonly used, one always has in mind the common, literal use, against which the eccentric, the metaphorical use, is counterpointed. One tends to make the ordinary inferences and this tendency is checked, so that there is always a sort of tension within a metaphor.
>
> A metaphor together with its context works like an assertion followed by a quick denial of something you would expect to follow from it . . . (McCloskey 1964:220–1)

The context of the metaphor (the animals in this case) arrests the usual habits we have of inferring with the literal use of the word. Thus, on seeing the metaphorical word 'greeted', all that we would normally associate with greeting is suddenly arrested on seeing its non-human context. However, it is not the denial that we remember, or that affects our reading of the metaphorical statement. Rather, all that we normally expect (in the literal use) of the metaphorical word remains in our minds. McCloskey noted:

> The implicit denial of a great deal of what the [metaphorical] word asserts does not make nearly the impression that its explicit apparent assertion makes, so that at least the flavour of all those things the word apparently asserts, lingers on after they have been withdrawn. (McCloskey 1964:221)

This is interesting in view of the fact that not one, but many statements in a primate behaviour report are so constructed (are metaphors), and there may be a cumulative effect on the reader to attribute the purposeful aspects of the metaphorical words to the animals.

As pointed out earlier, description of primate behaviour in terms of movement only, means something different from that described in ordinary language. Thus, the former sort of description is not a synonym for the latter, and some new meaning is created by the use of ordinary language terms. The notion of interaction-metaphor helps to explain this. The main point about the interaction view of metaphor is that the focus and the frame are active together and interact to produce a meaning that is a result of that interaction. The interesting outcome of this is that a change occurs in our concept of the principal subject (the chimpanzee, in our metaphor). The metaphorical word 'greeted' causes us to think of certain striking characteristics normally associated with its literal use (human greeting) and so reorganizes our view of the animals to which it is applied. To paraphrase an example that Black (1962) gives, all the human characteristics of greeting that can without undue strain be applied to chimpanzees, *will* be applied.

Earlier, a doubt was raised about whether the use of ordinary language in animal behaviour description makes a 'strong' metaphor. Of the different ways in which animal behaviour could be described, the fact that ordinary language description appears to be the most suitable raises the possibility that the terms can be used with their usual or literal meanings in describing animal behaviour, or that only a very weak metaphor has been created. What determines the strength or weight of a metaphor? Black remarked that:

> There are, in general, no standard rules for the degree of *weight* or *emphasis* to be attached to a particular use of an expression. To know what the user of a metaphor means, we need to known how 'seriously' he treats the metaphorical focus. (Would he be just as content to have a rough synonym, or would only *that* word serve? Are we to take the word lightly, attending only to its most obvious implications – or should we dwell upon its less immediate associations?) (Black 1962:29–30)

It seems that how seriously a metaphor is intended in descriptions of primate behaviour has changed and will continue to change along with scientists' attitudes or opinions as to what we can justifiably suppose goes on in an animal's mind. The proposers of cognitive ethology, for instance, would focus on mental phenomena in animals and probably tend to the literal rather than metaphorical use of ordinary language terms. In fact it is in just that sort of study where mental phenomena in animals are not assumed to be exactly like those of humans, that the terms may serve a catachretic function and thus become literal terms again.

The use of metaphor is in fact closely related to attitudes. Black (1962) noted that metaphorical language regularly produces shifts in attitudes by virtue of the metaphor reorganizing our view of the principal subject. Our acceptance of the literal application of ordinary language to animals may occur, if it does, partly through having had animals presented in a purposeful way in our ordinary language metaphors. A final point can be raised about metaphor with regard to the utility of anthropomorphism. Ordinary language terms can be heuristic aids to theory construction about the function of behaviour in primates.

8.10. Metaphor and theory construction in primate studies

Metaphor has often been assumed to be out of place in scientific terminology. McCloskey (1964) and Black (1962) held that metaphorical statements are too imprecise to be useful in sciences, mostly because a metaphor lacks a specific referent. As McCloskey explained, a metaphor calls to mind the striking characteristics of the literal use, but those characteristics do not serve as signs for other characteristics. One cannot make new inferences from metaphorical statements. Contrary to this view is that of Boyd (1979:357) who contends that metaphor can be and is used in science for the formation of theories:

> There exists an important class of metaphors which play a role in the development and articulation of theories in relatively mature sciences. Their function is a sort of catachresis – that is, they are used to introduce theoretical terminology where none previously existed. Nevertheless they possess several (though not all) of the characteristics which Black attributes to interaction metaphors; in particular, their success does not depend on their conveying quite specific respects of similarity or analogy. Indeed, their users are typically unable to precisely specify the relevant respects of similarity or analogy, and the utility of these metaphors in theory change crucially depends upon this open-endedness.
>
> (Boyd 1979:357)

Boyd noted that there are occasions when metaphorical expressions constitute, at least for a time, an irreplaceable part of the linguistic content of a scientific theory; cases where metaphors are used to express theoretical claims where there is no adequate literal paraphrase. These metaphors are constitutive of the theories they suggest, and Boyd calls them 'theory-constitutive metaphors'. Ordinary language terminology in primate ethology seems also to serve this catachretic function and to some extent directs formulation of theories about social behaviour.

The way in which Boyd's scheme applies to primate behaviour studies is as follows. Traditionally, most Western primate ethologists have not wished to imply subjective awareness in animals throughout their descriptions of them. Their use of an ordinary language terminology with purposeful connotations has raised the problem of anthropomorphism in those reports. That terminology is, however, the best one to use. It is the best because it gives the clearest and most understandable presentation of animal behaviour. These points have already been raised. To go a step further and ask why this presentation is clearest, brings us to the point that ordinary language description of animals allows us to formulate connections not only 'between' behaviours (i.e. to break up the stream of behaviour as between 'threaten' and 'appease'), but also to formulate more general theories about 'second-order' behaviour, such as 'strategies', 'preferences' (e.g. Bachmann & Kummer 1980) or social structure.

As mentioned, metaphorical terms do not correspond to precisely defined referents. Instead, in animal behaviour description, ordinary language changes 'behaviour' (O-purposive) into 'action' (A-purposive) and so generates a new universe of discourse about animal behaviour. Theories can then be (and are) generated about more complex aspects of social phenomena that may be occurring among the animals, such as reconciliation and consolation (de Waal & van Roosmalen 1979) or assessments by unattached males of *P. hamadryas* of females' 'preference' for a particular male before attempting to 'kidnap' her (Bachmann & Kummer 1980). These move the description of primate behaviour further into the realm of what we normally associate with human activities. This is inevitable, given the human context of the initial metaphor. As Boyd (1979:406) stated: 'Theory-constitutive metaphorical terms – when they refer – refer implicitly, in the sense that they do not correspond to explicit definitions of their referent, but instead indicate a research direction toward them'. In other words, the generic anthropomorphism or general purposefulness attributed to the animals' behaviour through metaphor is extended (as in words such as 'consolation') as more sophisticated analyses of social behaviour are made.

A final characteristic of metaphor that profoundly affects theory formation is that metaphor is culture-specific. Black stated:

> The important thing for the metaphor's effectiveness is that the system of associated commonplaces about the word used metaphorically are readily and freely evoked. Because this is so, a metaphor that works in one society may seem preposterous in another. (Black 1962:40)

McCloskey (1964) gave an example of a metaphor that would not travel: if someone calls heaps of fallen leaves 'breakfast food', it is understood by the nation that eats Kellogg's Cornflakes for breakfast, but not by a nation that eats porridge. The implications of this for the study of primate behaviour are considerable. What is associated by a particular observer or group with a key term, such as 'dominance' or 'troop' can affect what he looks for in the rest of the animal group's behaviour. For instance, the word 'troop' has connotations of simply an aggregation or assembly with no special structure or membership implied. It also can refer to soldiers or armed forces of people. For a time, some workers who used 'troop' carried over its militaristic connotations to specific aspects of social organization using words like 'marching order', 'van', 'sentinel' (e.g. Bolwig 1959; Washburn & DeVore 1961).

8.11. Quantitative description

It is sometimes suggested that use of quantitative description is a means of avoiding the anthropomorphism that attends qualitative description. In a very imprecise way, quantitative description is at times associated with 'objectivity' and qualitative description with 'subjectivity'. By the same token, quantification is vaguely apprehended to be more 'scientific' in its procedures and presentation of results. Against this trend, Konrad Lorenz remarked that:

> it has become fashionable to assess the so-called exactness and hence the value of any scientific result by the relative amount of mathematical operations used in obtaining it . . . quantification and mathematics have the last word in verification.
>
> (Lorenz 1975: 153)

Noel Cohen, a biologist, has called this attitude 'physics envy' and found it especially true of disciplines such as psychology, anthropology and sociology (Barash 1979:3).

These intuitions are, however, far from the truth of the matter. There is a place and a necessity for both kinds of description in primate ethology. Quantitative techniques have increasingly been applied and several aspects of quantitative description in primatology have been discussed in the literature, among them, kinds of recording devices (e.g. Hutt & Hutt 1970; Hinde 1973; Chamove 1974), sampling methods (e.g. J. Altmann 1974), statistical analysis of data (e.g. Slater 1973; Bramblett 1976) and some conceptual bases of its application to recording behaviour (e.g. Bramblett 1976; Colgan 1978, for ethology generally).

At times the qualitative dimension appears to be forgotten altogether. In the introduction to a chapter on techniques in primatology, Bramblett wrote:

> The scientific method begins with the rejection of authority. A statement is validated . . . by a demonstration of accuracy and predictability. In primatology the scientific approach usually means the analysis of statistical data based on observations of natural phenomena. Such observations are tabulated or quantified, and then analyzed in a standardized manner. (Bramblett 1976:261)

Before this can be done, however, qualitative definition of behaviour units is needed, as several primatologists have noted (e.g. Hall 1965;

Hinde 1973; Pruscha & Maurus 1973). Russell, Mead & Hayes (1954:199) remarked, long before these studies, that: 'The progress of the quantitative study of behaviour depends on the use of concepts which are qualitatively precise'. Even after quantitative methods have begun to be applied to the qualitative base Hinde (1973:402) noted that 'some aspects of behaviour will always be difficult to quantify, . . . [and] qualitative data may continue to be valuable throughout a quantitative study'.

Anthropomorphism arises at the level of behavioural categories (or complex behaviour units that can in turn be broken down into movement patterns) named in ordinary language terms. Behaviour may also be represented in notations such as numbers and the use of ordinary language terms apparently avoided. Numerical notation is not, however, semantically neutral. The ascription of number to behaviour units presupposes that the stream of behaviour has already been broken up into units and thus given a specific meaning. Furthermore, in the final analysis, such notations are retranslated back into ordinary language terms as are explanations of graphs and so on employed in the course of the study. A fundamental reason for this is that the data become more intelligible in ordinary language terms so this is a natural translation for the human mind to make. Additionally, although qualitative description is not as precise as quantitative description in one sense, in another sense it gives a more accurate picture of what occurs in a group. That is because the behaviour of each animal is unique and qualitative description is not bound by presentation of behaviour as an average of several varying events. Lorenz once explained this in an interview with R. I. Evans:

> We now know enough about the range of behavior patterns of normal geese to concentrate on the differences between individuals. . . . And you find that what emerges from this study is something like a very simplified model of characterology. For two years now, one of my students has been studying communication patterns among four geese and counting behavior patterns evident among them as they relate to each other. By making histograms of how often one goose threatens and another goose escapes, you find there are quantitative differences; although the behavior patterns themselves are all the same one goose may excel in bond behavior, and in fearlessness but be very weak in aggressivity. It's a placid goose. And it's a very likeable goose. To illustrate this, one goose has a very low threshold of both escape behavior and aggressive behavior. It's a funny combination – an aggressive coward – and this chap is what many normal persons would term vicious because he attacks suddenly from behind, bites and runs away and is very poor

in bond behavior. He is a loner but he breeds. This combination of quantitative differences gives you a very clearly described character of a goose, but describing the character of this goose is confoundedly anthropomorphic. (Evans 1975:10–11)

8.12. Conclusion

It has been my purpose to show that anthropomorphism in naturalistic primate behaviour reports arises partly by virtue of the nature of our language use. Description of primate behaviour was shown to proceed (at least implicitly) by identifying discrete movement patterns and vocalizations and giving them meaning by subsuming them under specific categories of behaviour according to principles chosen by the observer. The categories are often given in ordinary language terms such as 'threat', 'appease', 'dominance' and so on. It was contended that the everyday meaning of these ordinary language terms is not lost through the use of stipulated definitions based on behaviour units. Because ordinary language terms for social behaviour have most often been used to express human action, they connote purposefulness in the animals to which they are applied. This occurs through metaphor and gives rise to generic anthropomorphism, or the treatment of animals as conscious agents. The usual connections we make between purposeful actions serve as a heuristic aid to interpretation of the animals' behaviour. The sophistication and applicability of quantitative description was argued not to replace our ultimate understanding of animal behaviour in terms of ordinary language. It might be objected that the metaphorical use of ordinary language terms throughout a behaviour report does not give rise to anthropomorphism as long as the report is read in its biological (O-purposive, Purton 1970) context. As long as, for instance, 'appeasement' is discussed in terms of how the behaviour pattern aids the survival of the individual, how it compares with appeasement patterns in other species, how it evolved, and so on, we are out of the anthropomorphic realm. The notions of perception, purpose and motive are quite out of place here. It is only when the psychological realm (A-purposive, Purton 1970) is applied that the terms become anthropomorphic.

This study has suggested, firstly, that it is metaphor which makes it possible to apply the psychological context at all, and secondly, that this opportunity can prompt questions or generalizations about behaviour that are anthropomorphic. Examples are Simpson's (1973:416) set of questions: 'Does the frequency with which an individual chimpanzee grooms his fellows reflect "his preferences among" those individuals or is he often involved in grooming sessions because they "chose" him and perhaps even demanded grooming from him . . .?' or Hess's (1973:511) description of captive lowland gorillas (*Gorilla gorilla*): 'Later on Stefi usually *refrained* from entering the righthand cage, limiting himself to the middle and the lefthand one' (my emphasis). The purposeful connotations are always present in ordinary language vocabulary because the everyday meaning of the terms is not lost through the primatologists' stipulated definitions.

The source of the purposefulness is that ordinary language terms for social behaviour have most often been used to express human action. Nonetheless there appears to be only a vague notion of 'anthropomorphic licence' in primatologists' use of ordinary language terms. For instance, a term such as Hess's (1973) 'refrained' *seems* more anthropomorphic than a term such as 'greeted'. It is suggested here that these differences may be due to our familiarity with terms that commonly appear in primate behaviour reports. Terms such as 'threaten', 'submission', 'dominance', 'beg', 'chase', 'play', are so much a part of the literature that they do not jolt or surprise the reader. However, the fact that other terms, such as 'refrain' *do* surprise the reader is a reminder that the metaphorical process is constantly in the background and constantly implying the human reference of the terms.

The acceptance of many ordinary language terms in primate reports may not only be a matter of becoming used to their common recurrence. It may also be that primatologists have become practised in staying mentally within their stipulated definitions. Even so, terms read in an O-purposive context may get connected with more unusual terms such as 'reconciliation' and 'consolation' (de Waal & van Roosmalen 1979) and the links will be forged along the same lines as in human action (thus, 'appease' goes with 'threat'; 'submission' with 'dominance' and so on). In this way, the human referent serves as a heuristic aid to formation of social theories in primatology. This incidentally shows that generic anthropomorphism need not be misleading: on the contrary, the human action referent can provide a rich source of theories of social behaviour among primates.

Notes

1. See Asquith (1981) for a résumé of the effect of Christianity and Cartesianism on post-Baconian attitudes to animals.
2. For instance, Eisenberg, Muckenhirn & Rudran (1972) specify that their use of the term 'bond' refers only to social relationships between specific individuals based on the daily performance of mutually reinforcing activities (e.g. grooming, huddling) in addition to mating behaviours. A 'bond' is usually

associated with two individuals of the opposite sex and thus termed a 'pair bond'.

3. Several philosophers who have recently been interested in the distinction between action and movement are: Hamlyn (1953, 1964) Peters (1958), Melden (1961), Charles Taylor (1964), Richard Taylor (1966), White (1968) and Mischel (1969).
4. I am grateful to R. Harré for suggesting the notion of 'appropriateness' here.

References

Albrecht, H. & Dunnett, S. C. (1971). *Chimpanzees in Western Africa*. München: R. Piper & Co.

Altmann, J. (1974). Observational study of behavior: sampling methods. *Behaviour*, **49**, 227–67.

Altmann, S. A. (1962). A field study of the sociobiology of rhesus monkeys, *Macaca mulatta*. *Annals of the New York Academy of Sciences*, **102**, 338–435.

Asquith, P. J. (1978). A suggestion for fixed criteria definitions of terms that describe the social organization of non-human primates. In *Recent Advances in Primatology*, vol. 1, ed. D. J. Chivers & J. Nerbert, pp. 201–4, London: Academic Press.

Asquith, P. J. (1981). Some aspects of anthropomorphism in the terminology and philosophy underlying Western and Japanese studies of primate behaviour. D.Phil. thesis, University of Oxford.

Bachmann, C. & Kummer, H. (1980). Male assessment of female choice in hamadryas baboons. *Behavioural Ecology and Sociobiology*, **6**, 315–21.

Barash, D. (1979). *Sociobiology. The Whisperings Within*. New York: Harper & Row.

Black, M. (1962). *Models and Metaphors. Studies in Language and Philosophy*. Ithaca, New York: Cornell University Press.

Bolwig, N. (1959). A study of the behaviour of the chacma baboon. *Behaviour*, **14**, 136–63.

Box, H. O. & Pook, A. G. (1974). A quantitative method for studying behaviour in small groups of monkeys in captivity. *Primates*, **15**, 101–5.

Boyd, R. (1979). Metaphor and theory change: What is 'metaphor' a metaphor for? In *Metaphor and Thought*, ed. A. Ortony, pp. 356–408. Cambridge University Press.

Bramblett, C. A. (1976). *Patterns of Primate Behavior*. California: Mayfield.

Carpenter, C. R. (1942). Characteristics of social behavior in non-human primates. *Transactions of the New York Academy of Sciences*, **4**, 248–58.

Chamove, A. S. (1974). A new primate social behaviour category system. *Primates*, **15**, 85–99.

Colgan, P. W. (ed.) (1978). *Quantitative Ethology*. New York: Wiley.

Descartes, R. (1637). *Discourse on the Method of Rightly Directing One's Reason and of Seeking Truth in the Sciences*. Leyden: I. Maire. Translated from French by J. Veitch. Chicago: The Open Court Publishing Co.; London: K. Paul, Trench, Trübner & Co., 1910.

Eisenberg, J. F., Muckenhirn, N. A. & Rudran, R. (1972). The relation between ecology and social structure in primates. *Science*, **176**, 863–74.

Evans, R. I. (1975). *Konrad Lorenz. The Man and His Ideas*. New York & London: Harcourt Brace Jovanovich.

Gallup, G. R. (1977). Self-recognition in primates; a comparative approach to the bidirectional properties of consciousness. *American Psychologist*, **32**, 329–38.

Grant, E. C. (1969). Human facial expression. *Man*, **4**, 525–36.

Griffin, D. R. (1976). *The Question of Animal Awareness. Evolutionary Continuity of Mental Experience*. New York: Rockefeller University Press.
Griffin, D. R. (1978). Prospects for a cognitive ethology. *Behavioral and Brain Sciences*, **4**, 527–38.
Hall, K. R. L. (1965). Experiment and quantification in the study of baboon behavior in its natural environment. In *The Baboon in Medical Research*, ed. H. Vagtborg. pp. 29–42. Austin, Texas: University of Texas Press.
Hamlyn, D. W. (1953). Behaviour. *Philosophy*, **28**, 132–45.
Hamlyn, D. W. (1964). Causality and human behaviour. *Proceedings of the Aristotelian Society*, Supplementary Volume **38**, 125–45.
Harré, R. & Secord, P. F. (1972). *The Explanation of Social Behaviour*. Oxford: Basil Blackwell.
Hess, J. P. (1973). Some observations on the sexual behaviour of captive lowland gorillas (*Gorilla g. gorilla*, Savage and Wyman). In *Comparative Ecology and Behaviour of Primates*. ed. R. P. Michael & J. H. Crook, pp. 507–81. London: Academic Press.
Hinde, R. A. (1973). On the design of check-sheets. *Primates*, **14**, 393–406.
Hinde, R. A. (1979). The nature of social structure. In *The Great Apes* ed. D. A. Hamburg & E. R. McCown, pp. 295–315. Menlo Park, Calif.: Benjamin/Cummings Publishing Co.
Hopf, S. (1972). Study of spontaneous behaviour in squirrel monkey groups: observation techniques, recording devices numerical evaluation and reliability tests. *Folia Primatological*, **17**, 363–88.
Hopf, S., Hartmann-Wiesner, E., Kühlmorgen, B. & Mayer, S. (1974). The behavioral repertoire of the squirrel monkey (*Saimiri*). *Folia Primatologica*, **21**, 225–49.
Hornaday, W. T. (1922). *The Minds and Manners of Wild Animals*. New York: C. Scribner & Sons.
Hutt, S. J. & Hutt, C. (1970). *Direct Observation and Measurement of Behavior*. Springfield, III.: Charles C. Thomas.
Itani, J. (1961). The society of Japanese monkeys. *Japan Quarterly*, **8**, 421–30.
Jaynes, J. (1969). The historical origins of ethology and comparative psychology. *Animal Behaviour*, **17**, 601–6.
Katz, D. (1937). *Animals and Men: Studies in Comparative Psychology*. London: Longmans, Green.
Kummer, H. (1957). *Soziales Verhalten einer Mantelpaviangruppe*. Bern: Huber.
Lorenz, K. (1975). The fashionable fallacy of dispensing with description. In *Konrad Lorenz. The Man and His Ideas*, ed. R. I. Evans, pp. 152–80. New York & London: Harcourt Brace Jovanovich.
McCann, C. (1928). Notes on the common Indian langur (*Pithecus entellus*). *Journal of the Bombay Natural History Society*, **33**, 192–4.
McCloskey, M. A. (1964). Metaphors. *Mind*, **73**, 215–33.
Marler, P. R. & Hamilton, W. J. (1966). *Mechanisms of Animal Behavior*. New York: Wiley.
Martin, J. & Harré, R. (1982). Metaphor in science. In *Metaphor: Problems and Perspectives*, pp. 89–105, ed. D. Miall. Cambridge University Press.
Mason, W. A. (1979). Environmental models and mental modes: representational processes in the great apes. In *The Great Apes*, ed. D. A. Hamburg & E. R. McCown, pp. 277–93. Menlo Park, Calif.: Benjamin/Cummings Publishing Co.
Melden, A. I. (1961). *Free Action*. London: Routledge & Kegan Paul.
Menzel, E. W. Jr. (1975). Natural language of young chimpanzees. *New Scientist*, Jan. 16, 127–30.

Menzel, E. W. Jr (1978). Implications of chimpanzee language-training experiments for primate field research – and vice versa. In *Recent Advances in Primatology*, vol. 1, eds. D. J. Chivers & J. Herbert, pp. 883–95. London: Academic Press.

Midgley, M. (1978). *Beast and Man: The Roots of Human Nature.* Hassocks, Sussex: Harvester Press.

Mischel, T. (ed.) (1969). *Human Action: Conceptual and Empirical Issues.* New York: Academic Press.

Morris, D. (ed.) (1967). *Primate Ethology.* London: Weidenfeld & Nicolson.

Peters, R. S. (1958). *The Concept of Motivation.* London: Routledge & Kegan Paul.

Pruscha, H. & Maurus, M. (1973). A statistical method for the classification of behavior units occurring in primate communication. *Behavioral Biology*, **9**, 511–16.

Pruscha, H. & Maurus, M. (1976). The communicative function of some agonistic behaviour patterns in squirrel monkeys: the relevance of social context. *Behavioural Ecology and Sociobiology* **1**, 185–214.

Purton, A. C. (1970). Philosophical aspects of explanation in ethology. M. Phil. thesis, University of London.

Reynolds, V. (1967). *The Apes. The Gorilla, Chimpanzee, Orangutan and Gibbon – Their History and Their World.* London: Cassell.

Reynolds, V. (1976). The origins of a behavioural vocabulary: the case of the rhesus monkey. *Journal for the Theory of Social Behavior*, **6**, 105–42.

Reynolds, V. (1980). *The Biology of Human Action*, 2nd edn. Oxford & San Francisco: W. H. Freeman & Co.

Richards, I. A. (1936). *The Philosophy of Rhetoric.* Oxford University Press.

Robinson, R. (1972). *Definition.* Oxford: Clarendon Press.

Russell, W. M. S., Mead, A. P. & Hayes, J. S. (1954). A basis for the quantitative study of the structure of behaviour. *Behaviour*, **6**, 154–205.

Simonds, P. (1965). The Bonnet macaque in south India. In *Primate Behavior*, ed. I. De Vore, pp. 175–96. New York: Holt, Rinehart & Winston.

Simpson, M. J. A. (1973). The social grooming of male chimpanzees. In *Comparative Ecology and Behaviour of Primates*, ed. R. P. Michael & J. H. Crook. pp. 411–505. London: Academic Press.

Slater, P. J. B. (1973). Describing sequences of behavior. In *Perspectives in Ethology*, ed. P. P. G. Bateson & P. H. Klopfer. New York & London: Plenum Press.

Slater, P. J. B. (1978). Data collection. In *Quantitative Ethology*, ed. P. W. Colgan, pp. 8–24. New York: Wiley.

Smith, F. B. (1971). *Purpose in Animal Behaviour.* London: Hutchinson.

Strum, S. (1978). Dominance hierarchy and social organization. Strong or weak inference? In *Baboon Field Research: Myths and Models.* Wenner-Gren meeting, New York, June 25–July 4.

Taylor, C. (1964). *The Explanation of Behaviour.* London: Routledge & Kegan Paul.

Taylor, R. (1966). *Action and Purpose.* Englewood Cliffs, NJ: Prentice-Hall.

Thorpe, W. H. (1979). *The Origins and Rise of Ethology.* London: Heinemann Educational.

Tinbergen, N. (1951). *The Study of Instinct* (second printing with a new introduction, 1969). Oxford University Press.

Vaitl, E. (1978). Nature and implications of the complexly organised social systems in nonhuman primates. In *Recent Advances in Primatology*, vol. 1. ed. D. J. Chivers & J. Herbert, pp. 17–30. London: Academic Press.

Waal, F. B. M. de & Roosmalen, A. van (1979). Reconciliation and consolation among chimpanzees. *Behavioural Ecology and Sociobiology*, **5**, 55–66.

Washburn, S. L. & DeVore, I. (1961). The social life of baboons. *Scientific American*, **204**, (6), 62–71.

Weber, M. (1947). *The Theory of Social and Economic Organization*, translated by A. M. Henderson & Talcott Parsons, 2nd edn. London: Free Press.

Wheelwright, P. (1954). *The Burning Fountain. A Study in the Language of Symbolism*. Bloomington: Indiana University Press.

White, A. R. (1968). *The Philosophy of Action*. Oxford: Oxford University Press.

Wiepkema, P. R. (1961). An ethological analysis of the reproductive behaviour of the bitterling (*Rhodeus amarus* Bloch). *Archives Néerlandaise de Zoologie, Leiden*, **14**, 103–99.

COMMENT

ROY HARRIS

The point is well made by Asquith that we have no plausible alternative but to use an anthropomorphic conceptual framework in our analyses of animal communication. (That is not, incidentally, necessarily to agree with Asquith's attempt to construe the application of communicational vocabulary to animal signalling as being metaphorical.) As I see it, the trouble with interpreting the terms we use for describing animal communication by reference to the distinction 'literal' v. 'metaphorical' is the unfortunate implication that metaphorical assertions, descriptions, etc., are, precisely, not literally true, correct, etc. Whereas I would want to claim that many of these 'anthropomorphic' statements are indeed true or false in exactly the same way and by exactly the same criteria as parallel statements about human beings (e.g. 'X greeted Y', 'X threatened Y'). I can see no linguistic grounds for counting these as metaphors at all when applied to animals.

REPLY

PAMELA J. ASQUITH

If the definition of anthropomorphism as the ascription of specifically human traits to non-humans is used, then, in animal behaviour studies, the perennial question is whether or not psychological traits usually associated with humans should be applied to animals. In other words, is it sometimes useful to be anthropomorphic in describing animal behaviour? However, another question can be asked which I think lies at the basis of much of the present discussion. That is, '*Is* it anthropomorphic to describe animals in terms that imply purposeful behaviour?', or, in other words, 'Should the ordinary language terminology of human action be assumed to be literally applicable to humans only?' If the answer to that is 'no' at least some of the time then, when Harris says that he can see no linguistic grounds for counting anthropomorphic statements applied to animals as metaphors, I agree. It has little to do with linguistics *per se*, but rather with emotion, tradition, training or our opinion about whether or not we maintain the distinction between human and animal mind that anthropomorphism presupposes. I allude to this on p. 164 of the chapter. That is, 'is the use of ordinary language in animal behaviour studies a strong metaphor or are the terms sometimes used literally? How seriously a metaphor is intended depends on scientists' attitudes or opinions as to what we can justifiably suppose goes on in an animal's mind'.

However, to anyone who answers 'yes', he assumes ordinary language terms are literally applicable to humans only, and wishes to avoid attributing consciousness of this sort to animals, yet finds he has implied it in his use of ordinary language terminology, I would still suggest that the linguistic agent for the implied purposefulness is metaphor.

Thus I think that Harris's criticism is in fact directed to the assumption that the behaviour described by ordinary language terms is specific to humans. Anthropomorphism, by definition, presupposes a specificity to humans; human behaviour terms thus become metaphorical when applied in a non-human context.

REJOINDER 1

ROY HARRIS

My criticism, *pace* Dr Asquith, was *not* directed against the assumption that the behaviour described by ordinary language terms is specific to humans (although, in fact, I would not accept that assumption), but against the abuse of the term *metaphor*. A metaphor is not simply an expression which the speaker happens to believe is strictly inapplicable. *The ship of state foundered* is metaphorical in English, whereas *The sun rises in the east* is not: and that remains the case irrespective of my personal beliefs about political events or the orbit of the earth. There is no sense in which *This monkey threatened that monkey* can suddenly 'become' metaphorical when uttered by an ethologist who happens to believe that monkeys are purposeless automata.

REJOINDER 2

PAMELA J. ASQUITH

I am grateful to Professor Harris for pointing out that one's beliefs about the strict inapplicability of an expression does not 'make' it a metaphor.

Although the distinctive characteristics of genuine metaphorical efficacy (Black 1979) are by no means settled I agree with Professor Harris that ordinary language description of non-human primates is not strictly a metaphor. However, the process by which the stipulated definition of the ethogram is lost and generic anthropomorphism generated by use of ordinary language terms can usefully be described as a *metaphor-like process*. That is, various criteria by which we understand what it is that constitutes a metaphor (what happens when we use metaphor) are met by ordinary language description of animal behaviour and we can speak of a parallel to metaphor working as I have described in the paper. There is a precedent to speaking of a parallel to the metaphorical process in Kuhn (1979), where he derives how reference is established for 'natural-kind' (e.g. birds, sparrows, metals, heat, etc.) as distinct from theoretical ones.

There is little point in repeating the points of comparison with metaphor since calling the inevitable attribution of agency in primate behaviour described in ordinary language a metaphor-like process instead of a metaphor in no way alters the foregoing. An example may suffice: in metaphor, the metaphorical use of a word presupposes the literal one and the metaphorical word calls to mind the striking characteristics of the literal use. In animal behaviour description, although the ordinary language terms have been carefully defined in terms not implying a conscious agent, in the context of a report the usual connotation of agency slips in. We do not simply forget the stipulated definition, so why does this happen? Continuing the parallel, a metaphor is only appropriate (i.e. 'works') if the meaning of the metaphorical word can in some sense be associated with the meanings of the other literal words (in this case, animals) in the sentence. The appropriateness of ordinary language terms derives from how far we feel that the animals to which the behaviour is attributed are agents. Hence, the basis for the charge of anthropomorphism.

In this and other ways described in the paper a parallel to the metaphorical process can explain the generation of anthropomorphism in animal behaviour reports.

References

Black, M. (1979). More about metaphor. In *Metaphor and Thought*, ed. A. Ortony, pp. 19–43. Cambridge University Press

Kuhn, T. S. (1979). Metaphor in science. In *Metaphor and Thought*, ed. A. Ortony, pp. 409–19. Cambridge University Press.

Part III
Steps towards a solution

Introduction

'Action', used in the Weberian sense, is composed of 'behaviour', which is overt and directly observable, together with 'meaning', which is part of a semantic network and has to be inferred from the social context. Terrace analyses his own work, and that of others, on the language-learning skills of apes. He identifies grammar as an essential part of language. Grammar is the use of word-order rules to create meaning out of words. Despite the fact that Nim showed evidence of sign order, Terrace is skeptical that his utterances constitute language, since the possibility of rote-learning is not discounted, and it is puzzling that Nim's length of utterance did not increase over time as it does in children, and as one would expect it to if he were truly learning grammatical rules. In Terrace's view, therefore, the use of arbitrary signs in ape communication should not lead us to attribute human-like language to apes. While many signs may be learned by an ape, their use is insufficiently rule-bound to qualify as language. In the general context of the interpretation of primate signals, sign language studies on apes therefore give us two important pieces of information: first, that cognitive and intentional processes are very clearly at work in such cases, and second that there are clear limits, probably neural, to the extent to which cognitions and intentions can be put together to form complex utterances.

Reynolds likewise concludes that there are limits to the expression of intentionality in his study of rhesus monkeys. He shows that the behaviour of a number of individuals interacting with each other over a period of time seems to contain both intentional elements based on carefully structured cognitive strategies, and other elements that appear to be automatic and that could be innate behaviours selected during evolution. In so far as 'action' terminology can be applied to monkeys, he suggests that its meaning would have to be limited to a psychological rather than a sociological sense, because of the absence of a collective set of social rules in non-human primates. But because intentionality can be inferred in primate behaviour, meaning of some kind must be acknowledged, and therefore action terminology can be admitted.

Maurus & Ploog focus their attention on the outward form of social behaviour units and the problem of constructing the repertoire of a primate species (in this case squirrel monkeys). Leaving aside the question of intentionality, they show that there are distinct patterns in the communicative process. Using electrical stimulation as a tool to produce a standard social signal in one monkey, they record the ensuing responses of other monkeys and then, by means of sequence analysis and other methods, determine the species' ethogram for agonistic behaviour. Defining 'function' as the effects of a signal, they can thus generate a functional classification of behaviour and (by combining units with common effects), they can reduce the totality of units to a number of functional categories. The results of their studies fall very largely into the 'causal-automatic' frame of reference, and if cognitive-intentional elements are present in squirrel monkeys' social interactions they do not emerge from this type of analysis. As the authors point out, however, their studies provide a useful background for the study of meaning. They expose the structure of the communication system of the species, and further work will undoubtedly show how the behaviour of individuals in particular situations utilises this system in achieving particular personal objectives.

Chance & Jones lay emphasis on the existence in primates of a built-in tendency to stay together as members of a social group, which they see as an evolved tendency resulting from the danger of dispersion in the wild. This propensity underlies diverse social strategies that bring about cohesion. While they attribute the existence of these activities to natural selection they nevertheless emphasise the cognitive components of the behaviours concerned. Thus individuals are unlikely to behave inappropriately to each other after 'learning what to expect from others'. This learning process is continuous from infancy, beginning between mother and infant. Even at that stage, however, 'misunderstandings' can occur, and these may be important in establishing social competence in later life. While they do not, therefore, tackle the problem of 'meaning' directly, Chance & Jones clearly work with a model of social interaction in which an underlying substrate of evolved (therefore causal-automatic) processes are combined with a rich ontogenetic superstructure of learned components. These latter provide the basis for day-to-day activities, which are timed and shaped according to the knowledge individuals have about each other. Intentionality can then be inferred from the analysis of inter-individual behaviours over time.

A. Approaches to the interpretation of action

9. 'Language' in apes[1]

H. S. TERRACE

In 1917, Franz Kafka published a story about a chimpanzee who acquired the gift of human language. During the years which have elapsed since the publication of Kafka's tale, much has been written on the one hand about the presumably uniquely human capacity to use language and, on the other hand, about real apes whose human teachers claim that they have mastered certain features of human language. After attempting to teach my own chimpanzee to use a human language I questioned these claims on the very grounds that Kafka perceived to be the basis of his fictional chimpanzee's ability to talk:

> . . . there was no attraction for me in imitating human beings. I imitated them because I needed a way out, and for no other reason . . . And so I learned things, gentlemen. Ah, one learns when one needs a way out; one learns at all costs.
>
> (F. Kafka, *A Report to an Academy*)

In this chapter I will review the grounds for my skepticism about an ape's ability to learn a human language.

The question of what language is has yet to be answered by linguists, psychologists, psycholinguists, philosophers and other students of human language in a way that captures its many complexities in a simple definition. Agreement has been reached, however, about one basic property of all human languages. That is the ability to create new meanings, each appropriate to a particular context, through the application of grammatical rules. Noam Chomsky (1957) and George Miller (1964), among others, have convincingly reminded us of the futility of trying to explain a child's ability to create and understand sentences unless one attributes to the child a knowledge of rules that can generate an indeterminately large number of sentences from a finite vocabulary of words.

The dramatic reports of the Gardners, Premack, and Rumbaugh that a chimpanzee could learn substantial vocabularies of words of visual languages and that they were also capable of producing utterances containing two or more words (Gardner & Gardner 1969; Premack 1970; Rumbaugh, Gill & von Glaserfeld 1973), raise an obvious fundamental question: are a chimpanzee's multi-word utterances grammatical? In the case of the Gardners, one wants to know whether Washoe's signing *more drink* in order to obtain another cup of juice, or *water bird* upon seeing a swan, were creative juxtapositions of signs. Likewise one wants to know whether Sarah, Premack's main subject, was using a grammatical rule in arranging her plastic chips in the sequence *Mary give Sarah apple,* and whether Lana, the subject of a related study conducted by Rumbaugh, exhibited knowledge of a grammatical rule in producing the sequence *please machine give apple.*

In answering these questions, it is important to remember that a mere sequence of words does not qualify as a sentence. A rote-learned string of words presupposes no knowledge of the meanings of each element and certainly no knowledge of the relationships that exist between the elements. Sarah, for example, showed little (if any) evidence of understanding the meanings of *Mary, give,* and *Sarah* in the sequence *Mary give Sarah apple.* Likewise, it is doubtful that, in producing the sequence *please machine give apple,* Lana understood the meanings of *please, machine* and *give,* let alone the relationships between these symbols that would apply in actual sentences. There is evidence that Sarah or Lana could distinguish the symbol *apple* from symbols that named other reinforcers. This suggests that what Sarah and Lana learned was to produce rote sequences of the type ABCX, where A, B and C are nonsense symbols and X is a meaningful element. That conclusion is supported by the results of two studies: one an analysis of a corpus of Lana's utterances, the other an experiment on serial learning by pigeons.

Thompson & Church (1980) have shown that a major portion of Lana's utterances can be accounted for by three decision rules that dictate when one of six stock sentences might be combined with one of a small corpus of object or activity names. The decision rules are (*a*) did Lana want an ingestible object, (*b*) was the object in view, and (*c*) was the object in the machine? For example, if the object was in the machine, an appropriate stock sequence was *please move object name into machine,* and so on.

An experiment performed in my laboratory, showed that pigeons could learn to peck four simultaneously presented colors in a particular sequence (Straub *et al.* 1979; Straub & Terrace 1981). Such performance is of interest as evidence of the memorial capacity of pigeons. It does not, of course, justify interpreting the sequence of the colors A → B → C → D as the production of a sentence meaning *please machine give grain.*

While the sequences *please machine give grain* and *please machine give apple* are logically similar, they are not identical. It has yet to be shown that pigeons can learn ABCX sequences of the type that Sarah and Lana learned (where X stands for different reinforcers) or that pigeons can learn to produce different sequences for different reinforcers. But given the relative ease with which a pigeon can master an $A \rightarrow B \rightarrow C \rightarrow D$ sequence, neither of these problems seems that difficult, *a priori*. And even if a pigeon could not perform such sequences or, as would probably be the case, a pigeon learns them more slowly than a chimpanzee, we should not lose sight of the fact that learning a rote sequence does not require any ability to use a grammar.

Utterances of apes who were not explicitly trained to produce rote sequences pose different problems of interpretation. The Gardners report that Washoe was not required to sign sequences of signs nor was she differentially reinforced for particular combinations. She nevertheless signed utterances such as *more drink* and *water bird* (Gardner & Gardner 1974*a*, *b*). Before these and other utterances can be accepted as creative combinations of signs, combinations that create particular meanings, it is necessary to rule out simpler interpretations.

The simplest non-grammatical interpretation of such utterances is that they contain signs that are related solely by context. Upon being asked what she sees when looking in the direction of a swan it is appropriate for Washoe to sign *water* and *bird*. On this view, if Washoe knew the sign for sky, she might just as readily have signed such less interesting combinations as *sky water*, *bird sky*, *sky bird water*, and so on.

Even if one could rule out context as the only basis of Washoe's combinations, it remains to be shown that utterances such as *water bird* and *more drink* are constructions in which an adjective and a noun are combined so as to create a new meaning. In order to support that interpretation, it is necessary to show that she combined adjectives and nouns in a particular order. It is, of course, unimportant whether Washoe used the English order (adjective + noun) or the French order (noun + adjective) in creating combinations in which the meaning of a noun is qualified by an adjective. But it is important to show that presumed adjectives and nouns are combined in a consistent manner so as to create particular meanings.

It is, of course, true that sign order is but one of many grammatical devices used in sign language. Indeed sign order is less important in sign language than it is in spoken languages such as English. At the same time, sign order is one of the easiest, if not the easiest, grammatical devices of sign language to record. It also provides a basis for demonstrating an awareness of such simple constructions as subject–verb, adjective–noun, verb–object, subject–verb–object, and so on.

With only two minor exceptions, the Gardners have yet to publish any data on sign order as substantiated by a corpus of Washoe's combinations. Accordingly, the interpretation of combinations such as *water bird, more drink* remain ambiguous. One has too little information to judge whether such utterances are manifestations of a simple grammatical rule or whether they are merely sequences of contextually related signs.

9.1. Project Nim

One way of distinguishing between linguistic and simpler interpretations of an ape's signing is to examine a large body of the ape's utterances for regularities of sign order. The initial goal of Project Nim, which I started in 1973, was to amass and analyze such a corpus and thereby decide whether a chimpanzee could use one or more rules of a finite-state grammar. As puny as such an accomplishment might seem compared with language acquisition by a child, it would amount to a quantal leap in the linguistic ability of non-humans. As we shall see, showing that a chimpanzee can learn a mere finite-state grammar proved to be an elusive goal.

Socialization and training

The subject of our study was an infant male chimpanzee, named Nim Chimpsky. Nim was born at the Oklahoma Institute for Primate Studies in November, 1973 and was flown to New York at the age of 2 weeks. Until the age of 18 months he lived in the family home of a former student; subsequently he lived in a University-owned mansion in Riverdale where he was looked after by four students.

At the age of 9 months, Nim became the sole student in a small classroom complex I designed for him in the Psychology Department of Columbia University. The classroom allowed Nim's teachers to focus his attention more easily than they could at home and it also provided good opportunities for introducing Nim to many activities conducive to signing, such as looking at pictures, drawing and sorting objects. Another important feature of the classroom was the opportunity it provided for observing, filming and videotaping Nim without his being aware of the presence of visitors and observers who watched him through a one-way window or through cameras mounted in the wall of the classroom.

Nim's teachers kept careful records of what he signed both at home and in the classroom. During each session, the teacher dictated into a cassette recorder as much information as possible about Nim's signing and the context of that signing. Nim was also videotaped at home and in the classroom. A painstaking comparison of Nim's signing in both locales revealed no differences with respect to spontaneity, content, or any other feature. In view of comments attributed to other researchers of ape language, that Nim was conditioned in his nursery school like a rat or a pigeon in a Skinner box (cf. Bazar 1980), it should be emphasized that his teachers were just as playful and spontaneous in the Columbia classroom as they were at home. A detailed description of Nim's socialization and instruction in sign language has already been reported (Terrace 1979).

Nim's teachers communicated to him and amongst themselves in sign language. Although the signs Nim's teachers used were consistent with those of American Sign Language (ASL), their signing is best characterized as 'pidgin' sign language. This state of affairs is to be expected when a native speaker of English learns sign language. Inevitably, the word order of English superimposes itself on the teacher's signing, at the expense of the many spatial grammatical devices ASL employs. An ideal project would, of course, attempt to use only native or highly fluent signers as teachers, teachers who would communicate exclusively in ASL. There is, however, little reason to be concerned that this ideal has not been realized in Project Nim or for that matter, on *any* of the other projects which have attempted to teach sign language to apes. The achievements of an ape who truly learned pidgin sign language would be no less impressive than those of an ape who learned pure ASL. Both pidgin sign language and ASL are grammatically structured languages.

Nim was taught to sign by the methods developed by the Gardners (Gardner & Gardner 1969) and Fouts (1972): molding and imitation. During the 44 months he was in New York, he learned 125 signs, most of which were common and proper nouns (next frequent were verbs and adjectives, least frequent were pronouns and prepositions).

9.2. Combinations of two or more signs

Lexical regularities

During a two-year period, Nim's teachers recorded more than 20 000 of his utterances that consisted of two or more signs. Almost half of these utterances were two-sign combinations, of which 1378 were distinct. One characteristic of Nim's two-sign combinations led me to believe that they were primitive sentences. In many cases Nim used a particular sign in either the first or the second position, no matter what other sign it was combined with (Terrace 1979; Terrace *et al.* 1979). For example, *more* occurred in the first position in 85 percent of the two-sign utterances in which *more* appeared (such as *more banana, more drink, more hug,* and *more tickle*). Of the 348 two-sign combinations containing *give,* 78 percent had *give* in the first position. Of the 946 instances in which a transitive verb (such as *hug, tickle,* and *give*) was combined with *me* and *Nim,* 83 percent of them had the transitive verb in the first position.

These and other regularities in Nim's two-sign utterances are the first demonstrations I know of a reliable use of sign order by a chimpanzee. By themselves, however, they do not justify the conclusion that they were created according to grammatical rules. Nim could have simply imitated what his teachers were signing. That explanation seemed doubtful for a number of reasons. Nim's teachers had no reason to sign many of the combinations Nim had produced. Nim asked to be tickled long before he showed any interest in tickling; thus, there was no reason for the teacher to sign *tickle me* to Nim. Likewise, Nim requested various objects by signing *give* + *X* (*X* being whatever he wanted) long before he began to offer objects to his teachers. More generally, all of Nim's teachers and many experts on child language learning, some of whom knew sign language, had the clear impression that Nim's utterances typically contained signs that were not imitative of the teacher's signs.

Another explanation of the regularities of Nim's two-sign combinations that did not require the postulation of grammatical competence was statistical. However, an extensive analysis of the regularities observed in Nim's two-sign combinations showed that they did not result from Nim's preferences for using particular signs in the first or second positions of two-sign combinations. Finally the sheer variety and number of Nim's combinations makes implausible the hypothesis that he somehow memorized them (Terrace *et al.* 1980).

Semantic relationships expressed in Nim's two-sign combinations

Semantic distributions, unlike the lexical ones discussed above, cannot be constructed directly from a corpus. In order to derive a semantic distribution, observers have to make judgments as to what each combination means. Procedures for making such judgments, introduced by Bloom (1970, 1973) and Schlesinger (1971), are known as the method of 'rich interpretation' (Brown 1973; Fodor, Bever & Garrett 1974; Bloom & Lahey 1977). An observer relates the utterance's immediate context to its contents. Supporting evidence for semantic judgments includes the following observations. The child's choice of word order is usually the same as it would be if the idea were being expressed in the canonical adult form. As the child's mean length of utterance (MLU) increases, semantic relationships identified by a rich interpretation develop in an orderly fashion (Bloom 1973; Bowerman 1973; Brown 1973). The relationships expressed in two-word combinations are the first ones to appear in the three- and four-word combinations. Many longer utterances appear to be composites of the semantic relationships expressed in shorter utterances (Bloom 1973; Brown 1973).

Studies by the Gardners and by Patterson of an ape's ability to express semantic relationships in combinations of signs have yet to advance beyond the stage of unvalidated interpretation. The Gardners (Gardner & Gardner 1971) and Patterson (1978) concluded that a substantial portion of Washoe's and Koko's two-sign combinations were interpretable in categories similar to those used to describe two-word utterances of children (78 and 95 percent of Washoe's and Koko's two-sign combinations, respectively). No data are available as to the reliability of the interpretations that the Gardners and Patterson have advanced.

Without prejudging whether Nim actually expressed semantic relationships in his combinations, we analyzed, by the method of rich interpretation, 1262 of his two-sign combinations, which occurred between the ages of 25 and 31 months. Twenty categories of semantic relationships account for 895 (85 percent) of the 957 interpretable two-sign combinations. Brown (1973) found that there were 11 semantic relationships that account for about 75 percent of all combinations of the children he studied. Similar categories of semantic relationships were used by the Gardners (Gardner & Gardner 1971) and by Patterson (1978).

Figure 9.1 shows several instances of significant preferences for placing signs expressing a particular semantic role in either the first or the second positions. Agent, attribute, and recurrence (*more*) were expressed most frequently in the first position, while place and beneficiary roles were expressed most frequently by second-position signs.

9.3. Differences between Nim's and a child's combinations of signs and words

The analyses performed on Nim's combinations of signs provided the most compelling evidence I know of that a chimpanzee could use grammatical rules, albeit finite-state rules, for generating two-sign sequences. It was not until after our funds ran out and it became

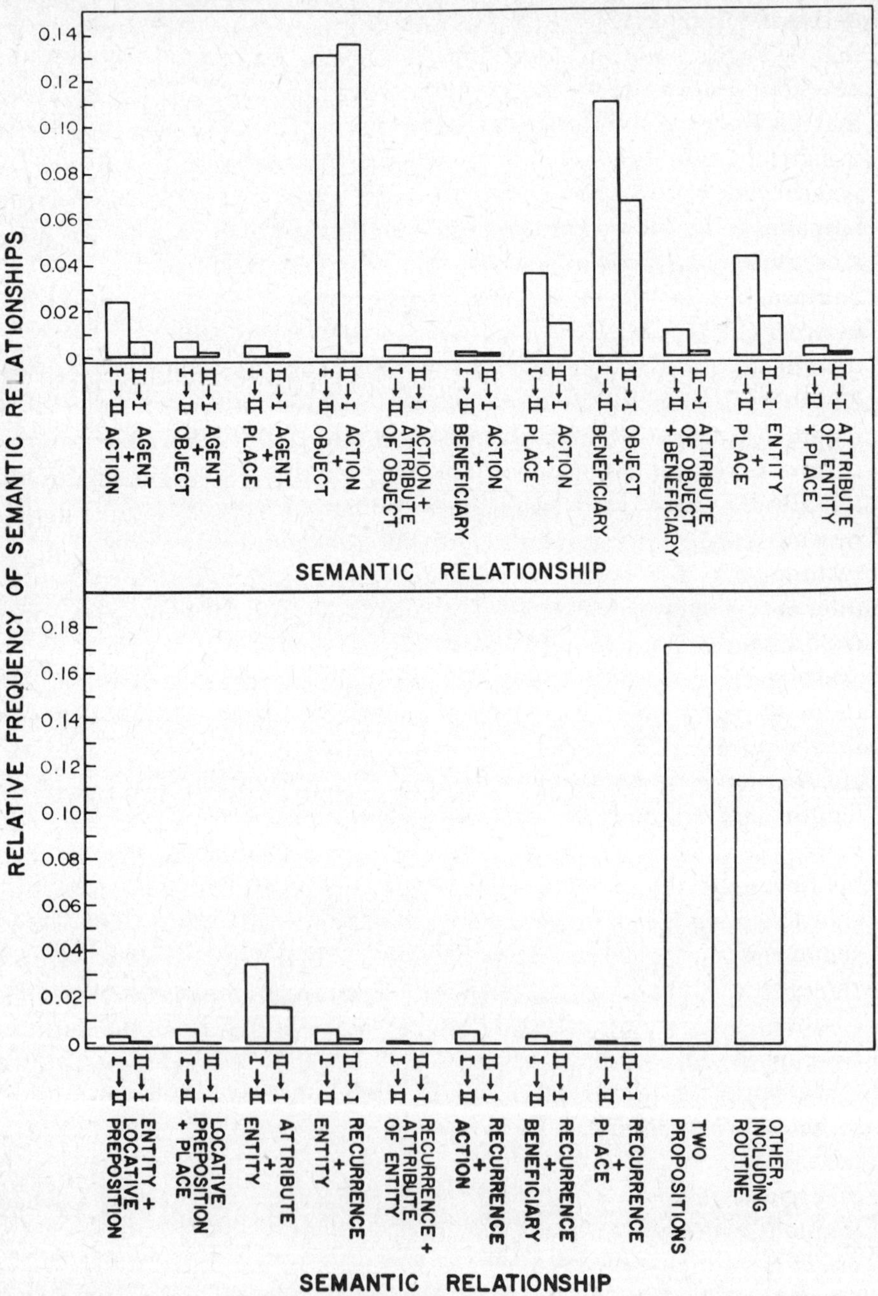

Figure 9.1. Relative frequencies of different semantic relationships. The bars above I → II show the relative frequencies of two-sign combinations expressing the relationship in the order specified under the bar: for example, an agent followed by an action. The bars above II→I show the relative frequencies of two-sign combinations expressing the same relationship in the reverse order: for example, an action followed by an agent.

necessary to return Nim to the Oklahoma Institute for Primate Studies that I became skeptical of that conclusion. Ironically, it was our new-found freedom from data-collecting, teaching and looking after Nim that allowed me and other members of the project to examine Nim's use of sign language more thoroughly. What emerged from our new analyses was a number of important differences between Nim's and a child's use of language. One of the first facts that troubled me was the absence of any increase in the length of Nim's utterances. During the last two years that Nim was in New York, the average length of Nim's utterances fluctuated between 1.1 and 1.6 signs. That performance is similar to what children do when they begin combining words. Furthermore, the maximum length of a child's utterances is related very reliably to their average length. Nim's showed no such relationship.

As children get older, the average length of their utterances increases steadily. As shown in Figure 9.2, this is true both of children with normal hearing and of deaf children who sign. After learning to make utterances relating a verb and an object (as, for example, *eats breakfast*), and utterances relating a subject and a verb (as, for example, *Daddy eats*), the child learns to link them into longer utterances relating the subject, verb, and object (for example, *Daddy eats breakfast*). Later, the child learns to link them into longer utterances such as *Daddy didn't eat breakfast*, or *when will Daddy eat breakfast?*

Despite the steady increase in the size of Nim's vocabulary, the mean length of his utterances did not increase. Although some of his utterances were very long, they were not very informative. Consider, for example, his longest utterance, which contained 16 signs: *give orange me give eat orange me eat orange give me eat orange give me you*. The same kinds of run-on sequences can be seen in comparing Nim's two-, three- and four-sign combinations. As shown in Table 9.1, the topic of Nim's three-sign combinations overlapped considerably with the apparent topic of his two-sign combinations.

Eighteen of Nim's 25 most frequent two-sign combinations can be seen in his 25 most frequent three-sign combinations, in virtually the same order in which they appear in his two-sign combinations. Furthermore, if one ignores sign order, all but five signs that appear in Nim's 25 most frequent two-sign combinations (*gum, tea, sorry, in,* and *pants*) appear in his 25 most frequent three-sign combinations. We did not have enough contextual information to perform a semantic analysis of Nim's two- and three-sign combinations. However, Nim's teachers' reports indicate that the individual signs of his combinations were appropriate to their context and that equivalent two- and three-sign combinations occurred in the same context.

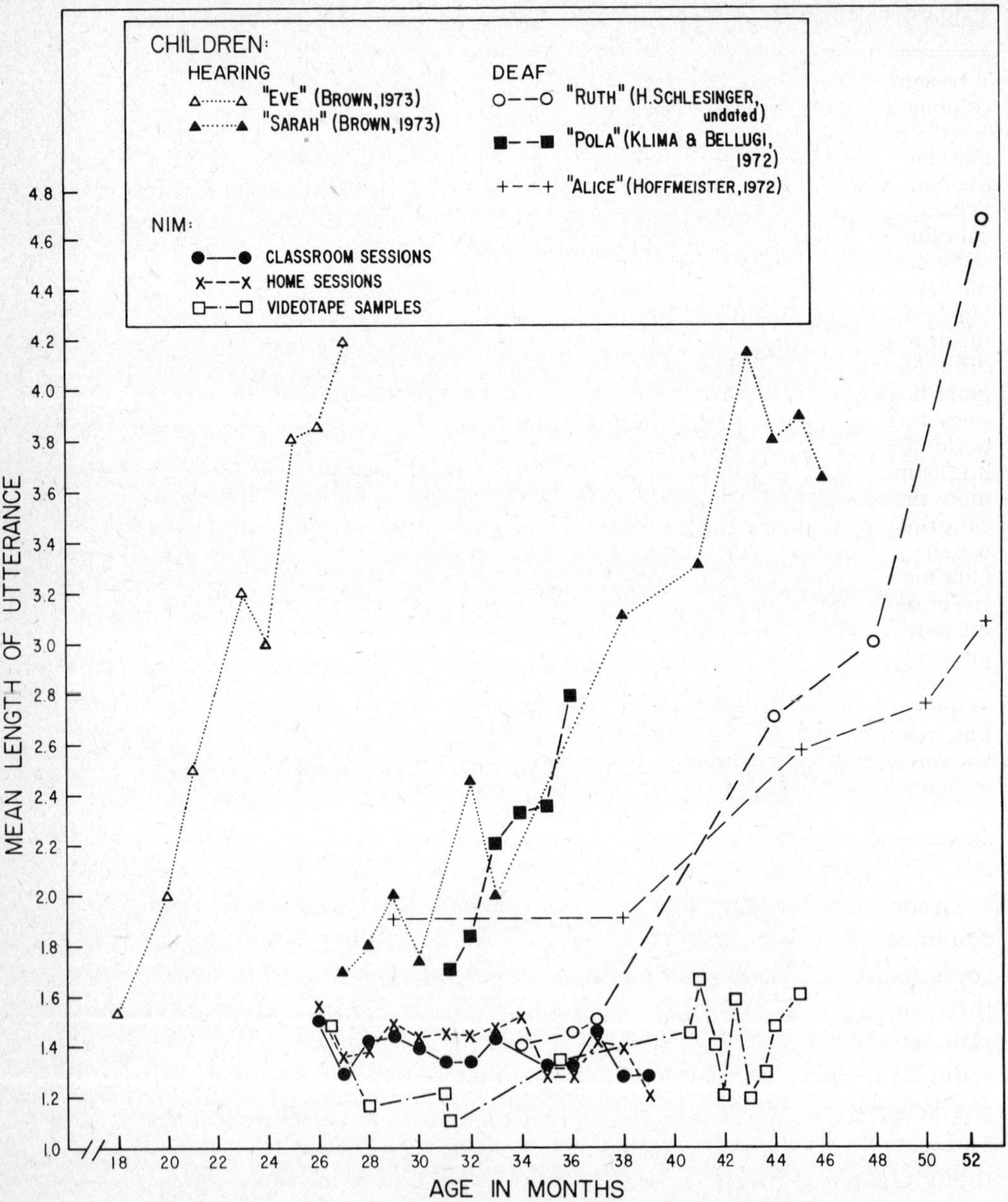

Figure 9.2. Mean length of signed utterances (MLU) of Nim and three deaf children and mean length of spoken utterances of two hearing children. The functions showing Nim's MLU between January 1976 and February 1977 (age 29–39 months) are based on data obtained from teachers' reports; the function showing Nim's MLU between February 1976 and August 1977 (age 27–45 months) is based upon video-transcript data.

Table 9.1. *Most frequent two- and three-sign combinations*

Two-sign combinations	Frequency	Three-sign combinations	Frequency
play me	375	play me Nim	81
me Nim	328	eat me Nim	48
tickle me	316	eat Nim eat	46
eat Nim	302	tickle me Nim	44
more eat	287	grape eat Nim	37
me eat	237	banana Nim eat	33
Nim eat	209	Nim me eat	27
finish hug	187	banana eat Nim	26
drink Nim	143	eat me eat	22
more tickle	136	me Nim eat	21
sorry hug	123	hug me Nim	20
tickle Nim	107	yogurt Nim eat	20
hug Nim	106	me more eat	19
more drink	99	more eat Nim	19
eat drink	98	finish hug Nim	18
banana me	97	banana me eat	17
Nim me	89	Nim eat Nim	17
sweet Nim	85	tickle me tickle	17
me play	81	apple me eat	15
gum eat	79	eat Nim me	15
tea drink	77	give me eat	15
grape eat	74	nut Nim nut	15
hug me	74	drink me Nim	14
banana Nim	73	hug Nim hug	14
in pants	70	play me play	14
		sweet Nim sweet	14

Though lexically similar to two-sign combinations, the three-sign combinations do not appear to be informative elaborations of two-sign combinations. Consider, for example, Nim's most frequent two- and three-sign combinations: *play me* and *play me Nim*. Combining *Nim* with *play me* to produce the three-sign combination, *play me Nim*, adds a redundant proper noun to a personal pronoun. Repetition is another characteristic of Nim's three-sign combinations, for example, *eat Nim eat*, and *nut Nim nut*. In producing a three-sign combination, it appears as though Nim is adding to what he might sign in a two-sign combination, not so much to add new information, but instead to add emphasis. Nim's most frequent four-sign combinations reveal a similar picture. In children's utterances, by contrast, the repetition of a word, or a sequence of words, is a rare event.

Having seen what Nim signed about in two-, three- and four-sign combinations it is instructive to see what he signed about with single signs. As shown in Table 9.2, the topics of Nim's most frequent 24 single-sign utterances overlap considerably with those of his most frequent multi-sign utterance. (Most of the exceptions are signs required in certain routines, e.g. *finish*, when Nim was finished using the toilet; *sorry*, when Nim was scolded, and so on.) In contrast to a child, whose longer utterances are semantically and syntactically more complex than his shorter utterances, Nim's are not. When signing a combination, as opposed to signing a single sign, Nim gives the impression that he is running on with his hands. It appears that Nim learned that the more he signed, the better his chances of obtaining what he wanted. It also appeared as though Nim made no effort to add informative, as opposed to redundant, signs in satisfying his teacher's demand that he sign.

At first glance, the results of our semantic analysis appear to be consistent with the observations of the Gardners and Patterson. But even though I could demonstrate the reliability of our judgments, several features of our results suggest that our analysis, and that of others, may exaggerate an ape's semantic competence. One problem is the subjective nature of semantic interpretations. That problem can be remedied only to the extent that evidence corroborating the psychological reality of our interpretations is available. Neither our semantic analyses of Nim's two-sign combinations nor those of any other studies have produced such evidence. In some cases, utterances were inherently equivocal in our records. Accordingly, somewhat arbitrary rules were used to interpret these utterances. Consider, for example, combinations of *Nim* and *me* with an object name (for example, *Nim banana*). These occurred when the teacher held up an object that the teacher was about to give to Nim who, in turn, would ingest it. We had no clear basis for distinguishing between the following semantic interpretations of combinations containing *Nim* or *me* and an object name: agent–object, beneficiary–object, and possessor–possessed–object.

An equally serious problem is posed by the very small number of lexical items used to express particular semantic roles. Only when a semantic role is represented by a large variety of signs is it reasonable to attribute position preferences to semantic rules rather than to lexical position habits. For example, the role of recurrence was presented exclusively by *more*. In combinations presumed to relate an agent and an object, or an object and a beneficiary, one would expect agents and beneficiaries to be expressed by a broad range of agents and beneficiaries: for example, *Nim*, *me*, *you*, and names of other animate beings. However, 99 percent (N = 297) of the beneficiaries in utterances judged to be object–beneficiary combinations were *Nim* and *me*, and 76 percent (N = 35) of the agents in

Table 9.2. *Twenty-four most frequent single signs July 5, 1976–February 7, 1977*

Sign	Tokens
hug	1650
play	1545
finish	1103
eat	951
dirty	788
drink	712
out	615
Nim	613
open	554
tickle	414
bite	407
shoe	405
red	380
pants	372
sorry	366
angry	354
me	351
banana	348
nut	323
down	316
toothbrush	302
more	301
grape	239
sweet	236

utterances judged to be agent–object combinations were *you*. From these and other examples, it is difficult to decide whether the positional regularities favoring agent–object and object–beneficiary constructions (Figure 9.1) are expressions of semantic relationships or idiosyncratic lexical position habits. Such isolated effects could also be expected to appear from statistically random variation.

Discourse analysis

The most dramatic difference between Nim's and a child's use of language was revealed in a painstaking analysis of videotapes of Nim's and his teacher's signing. These tapes revealed much about the nature of Nim's signing that could not have been perceived at the time. Indeed, they were so rich information that it took as much as one hour to transcribe a single minute of tape.

Terrace *et al.* (1979) showed that Nim's signing with his teachers bore only a superficial resemblance to a child's conversations with his or her parents. What is more, only 12 percent of Nim's utterances were not preceded by a teacher's utterance. A significantly larger proportion of a child's utterances is spontaneous. In addition to differences in spontaneity, there were differences in creativity.

As a child gets older, the proportion of utterances that are full or partial imitations of the parent's prior utterance(s) decreases from less than 20 percent at 21 months to almost zero by the time the child is three years old. When Nim was 26 months old, 38 percent of his utterances were full or partial imitations of his teacher's. By the time he was 44 months old, the proportion had risen to 54 percent. A doctoral dissertation by Richard Sanders (Sanders 1980) showed that the functions of Nim's imitative utterances differed from those of a child's imitative utterances. Bloom Rocissano & Hood (1976) have observed that children imitated their parents' utterances mainly when they were learning new words or new syntactic structures. Sanders found no evidence for either of these functions in Nim's imitative utterances.

As children imitate fewer of their parents' utterances, they begin to *expand* upon what they hear their parents say. At 21 months, 22 percent of a child's utterances add at least one word to the parent's prior utterance; at 36 months, 42 percent are expansions of the parent's prior utterance. Fewer than 10 percent of Nim's utterances recorded during 22 months of videotaping (the last 22 months of the project) were expansions. Like the mean length of his utterances, this value remained fairly constant.

The videotapes brought out another distinctive feature of Nim's conversations. He was as likely to interrupt his teacher's signing as not. In contrast, children interrupt their parents so rarely (so long as no other speakers are present) that interruptions are all but ignored in studies of their language development. A child learns readily what one takes for granted in a two-way conversation: each speaker adds information to the preceding utterance and each speaker takes a turn in holding the floor. Nim rarely added information and showed no evidence of turn-taking.

None of the features of Nim's discourse – his lack of spontaneity, his partial imitation of his teacher's signing, his tendency to interrupt – had been noticed by any of his teachers or by the many expert observers who had watched Nim sign. Once I was sure that Nim was not imitating *precisely* what his teacher had just signed, I felt that it was less important to record the teachers' signs than it was to capture as much as I could about Nim's signing: the context and specific physical movements, which hand he signed with, the order of his signs, and their appropriateness. But even if one wanted to record the teacher's signs, limitations of attention span would make it too difficult to remember all of the significant features of *both* the teacher's and Nim's signs.

Once I knew what to look for, the contribution of the teacher was easy to see, embarrassingly enough in still photographs that I had looked at for years. Consider, for example, Nim's signing the sequence *me hug cat* as shown in Figure 9.3. At first, these photographs (and many others) seemed to provide clear examples of spontaneous and meaningful combinations. But just as analyses of our videotapes provided evidence of a relationship between Nim's signing and the teacher's prior signing, a re-examination of Figure 9.3 revealed a previously unnoticed contribution of the teacher's signing. She signed *you* while Nim was signing *me*, and signed *who* while he was signing *cat*. While Nim was signing *hug*, his teacher held her hand in the 'n'-hand configuration, a prompt for the sign *Nim*. Because these were the only photographs taken of this sequence, we cannot specify just when the teacher began her signs. It is not clear, for example, whether the teacher signed *you* simultaneously or immediately prior to Nim's *me*. It is, however, unlikely that the teacher signed *who?*

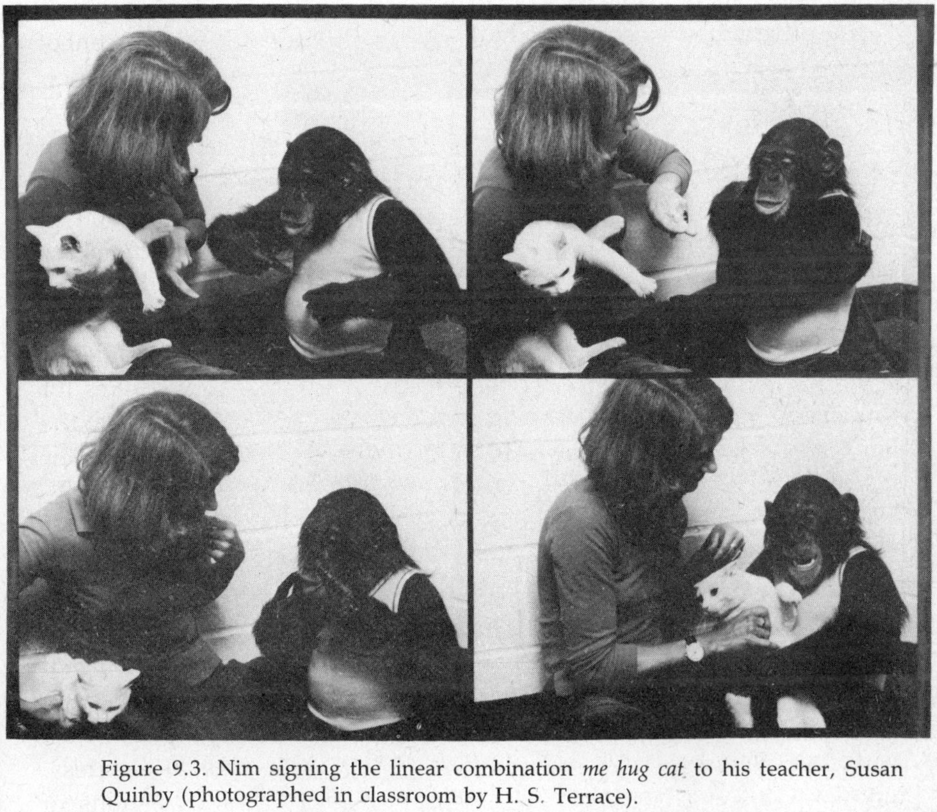

Figure 9.3. Nim signing the linear combination *me hug cat* to his teacher, Susan Quinby (photographed in classroom by H. S. Terrace).

after Nim signed *cat*. A few moments before these photographs were taken the teacher repeatedly quizzed Nim as to the contents of the cat box by signing *who?* At the very least, Nim's sequence, *me hug cat*, cannot be interpreted as a spontaneous combination of three signs.

There remain other differences between Nim's and a child's signing that need to be explored. Two important questions are how similar the contents of a chimpanzee's and a child's utterances are, and whether or not they are structured grammatically. A cursory examination of corpora of children's utterances indicates that chimpanzees and children differ markedly with respect to the variety of the utterances they make. While children produce certain routine combinations, they are relatively infrequent. Nim's utterances, on the other hand, show a high frequency of routine combinations (e.g. *play me, me Nim;* see Table 9.1 for additional examples).

It also remains to be shown that the patterns of vocal discourse observed between a hearing parent and a hearing child are similar to the patterns of signing discourse that obtain between a deaf parent and a deaf child. Videotapes of such discourse are now available from a number of sources (Bellugi & Klima 1976; Hoffmeister 1978; McIntire, 1978).

9.4. Criticisms of Project Nim

Of more immediate concern is the generality of the conclusions drawn about Nim's signing. This issue can be approached in two ways. One is to consider the methodological weaknesses of Project Nim and to pursue their implications. The other is to ignore Project Nim and to ask whether other signing apes sign because they are coaxed to do so by the teachers and how much overlap exists between the teacher's and the ape's signing.

Consider first some of the initial questions raised about Project Nim. It has been said that Nim was taught by too many teachers (60 all told), that his teachers were not fluent enough in ASL, that terminating Nim's training at the age of 44 months prevented his teachers from developing Nim's full linguistic competence and that Nim may simply have been a stupid chimpanzee (e.g., quotes of other researchers studying language in apes as reported by Bazar 1980).

Aside from the speculation that Nim may have been a stupid chimp, I believe that all of these criticisms are valid. And, if questioning Nim's intelligence is simply a nasty way of asking whether a sample size of one is an adequate basis for forming a general conclusion about an ape's grammatical competence, I would readily admit that it is not. A case can be made, however, that most of the methodological inadequacies of Project Nim have been exaggerated and, in any event, that they are hardly unique to Project Nim. Though Nim was taught by 60 teachers he spent most of his time in the presence of a core group of 8 teachers: Stephanie LaFarge, Laura Petitto, Amy Schacther, Walter Benesch, Bill Tynan, Joyce Butler, Dick Sanders, and myself. As already described (Terrace 1979, see especially Appendix B), many of Nim's 60 teachers served as occasional playmates rather than as regular teachers. Nevertheless, they were each listed as a teacher. The Gardners have yet to publish a full list of the teachers who worked with Washoe. Roger Fouts, however, estimates that their number was approximately 40. Of greater importance is Fouts' observation that, during Washoe's four years in Reno, Nevada, she was looked after mainly by a small group of six core teachers.

Francine Patterson's doctoral dissertation presents data provided by 20 teachers during the three-year period in which they worked with Koko (Patterson 1979). This is hardly surprising. It was difficult for an ape's teachers to sustain the energy needed to carry out lesson plans, to engage its attention, to stimulate it to sign, and to record what it signed for more than 3–4 hours a day. A 3–4 hour session with Nim also entailed an additional hour or two for transcribing the audio cassette on which the teacher dictated information about Nim's signing and writing a report. At least six full-time people would be needed to carry out such a schedule on a 16-hour per day, 7-day per week basis. As far as I know, no project has been able to afford the salaries for such a staff. Accordingly, it is necessary to make do with a large contingent of part-time volunteers.

Both Patterson and Fouts speak English while signing with their apes. Films of the Gardners and Fouts signing with Washoe, of Fouts signing with Ally and Booee (two resident chimpanzees of the Oklahoma Center for Primate Research), and of Francine Patterson signing with Koko make clear that none of these researchers uses ASL. As mentioned earlier, pidgin sign seems to be the prevalent form of communication on *all* projects attempting to teach apes to use sign language.

Washoe is now 15, Koko is 9, and Ally is 9 years old. I know of no evidence that their linguistic skills increased as they became older. An ape's intelligence undoubtedly increases after infancy. One must, however, also keep in mind that as an ape gets older, its ability to master its environment by physical means also increases. An ape's increasing strength and its recognition that it can get its way without signing would make the teacher less dominant. As a result the ape would be *less* motivated to sign. I am therefore skeptical of conjectures that an ape's increasing intelligence would manifest itself in a more sophisticated use of language.[2]

9.5. The generality of the conclusions of Project Nim

Whatever the shortcomings of Project Nim it should be recognized that they are irrelevant to the following hypothesis about an ape's use of signs. An ape signs mainly in response to his teacher's urgings, in order to obtain certain objects or activities. Combinations of signs are not used creatively to generate particular meanings. Instead, they are used for emphasis or in response to the teacher's unwitting demands that the ape produce as many contextually relevant signs as possible. The validity of this hypothesis rests simply on the nature of data obtained from other signing apes.

Finding such data has proven difficult, particularly because discourse analyses of other signing apes have yet to be published. Also, as mentioned earlier, published accounts of an ape's combinations of signs have centered around anecdotes and not around exhaustive listings of all combinations. One can, however, obtain some insight into the nature of signing by other apes by looking at films and videotape transcripts of their signing. Two films made by the Gardners of Washoe's signing, a doctoral dissertation by Lyn Miles (1978; which contains four videotape transcripts of two Oklahoma chimps, Ally and Booee), and a recently released film *Koko, the Talking Gorilla,* all support the hypothesis that the teacher's coaxing and cueing have played much greater roles in so-called 'conversations' with chimpanzees than was previously recognized.

In a 'Nova'-produced film, *The First Signs of Washoe,* Beatrice Gardner can be seen signing *what time now?*, an utterance that Washoe interrupts to sign *time eat, time eat.* A longer version of the same exchange shown in the second film, *Teaching Sign Language to the Chimpanzee Washoe,* began with B. T. Gardner signing *eat me, more me,* after which Washoe gave Gardner something to eat. Then she signed *thank you* and asked *what time now?* Washoe's response *time eat, time eat* can hardly be considered spontaneous, since Gardner had just used the same signs and Washoe was offering a direct answer to her question.

The potential for misinterpreting an ape's signing because of inadequate reporting is made plain by another example in both films. Washoe is conversing with her teacher, Susan Nichols, who shows the chimp a tiny doll in a cup. Nichols points to the cup and signs *that;* Washoe signs *baby.* Nichols brings the cup and doll closer to Washoe, allowing her to touch them, slowly pulls them away, and then signs *that* by pointing to the cup. Washoe signs *in* and looks away. Nichols brings the cup and doll closer to Washoe again, who looks at the two objects once more, and signs *baby*. Then, as she brings the cup still closer, Washoe signs *in*. *That*, signs Nichols, and points to the cup; *my drink*, signs Washoe.

Given these facts, there is no basis for referring to Washoe's utterance – *baby in baby in my drink* – as a spontaneous or a creative use of 'in' as a preposition joining two objects. It is actually a 'run-on' sequence with very little relationships between its parts. Only the last two signs were uttered without prompting from the teacher. Moreover, the sequence of the prompts (pointing to the doll, and then pointing to the cup) follows the order called for in constructing an English prepositional phrase. In short, discourse analysis makes Washoe's achievement less remarkable than it might seem at first.

In commenting about the fact that Koko's MLU was low in comparison with that of both hearing and deaf children, Francine Patterson speculates, in her doctoral dissertation (Patterson 1979:153), that 'This probably reflects a species difference in syntactic and/or sequential processing abilities'. Patterson goes on to observe that 'The majority of Koko's utterances were not spontaneous, but elicited by questions from her teachers and companions. My interactions with Koko were often characterized by frequent questions such as "What's this?"' (Patterson 1979:153).

Four transcripts appended to Lyn Miles' dissertation provided me with a basis for performing a discourse analysis of the signing of two other chimpanzees, Ally and Booee. Each transcript presents an exhaustive account of one of these chimps signing with one of two trainers: Roger Fouts and Joe Couch. The MLU and summaries of the discourse analysis of each tape is shown in Table 9.3. Tables 9.4 and 9.5 show exhaustive summaries of two conversations, one from a session with the highest MLU, and one from a session with the highest percentage of novel adjacent utterances. The novel utterances are very similar to Nim's run-on sequences. They also overlap considerably with adjacent utterances that were expansions and with non-contingent utterances.

In his discussion of communicating with an animal, the philosopher Ludwig Wittgenstein cautions that apparent instances of an animal using human language may prove to be a 'game' that is played by simpler rules. Nim's, Washoe's, Ally's, Booee's, and Koko's use of signs suggests a type of interaction between an ape and its trainer that has little to do with human language. In each instance the sole function of the ape's signing appears to be to request various rewards that can be obtained only by

Table 9.3. *Summary of Ally and Booee transcripts*

Videotape number	3	4	5	6	Mean
Number of utterances	38	79	102	72	72.75
MLU	1.63	1.52	2.25	1.93	1.85
Adjacent (%)	76.3	93.7	77.4	86.1	83.4
Imitations (%)	13.6	22.8	7.84	8.33	13.03
Expansions (%)	7.89	7.59	13.7	4.16	8.34
Novel (%)	55.3	63.3	55.9	73.6	62.3
Non-contingent (%)	23.7	6.3	2.6	13.9	16.6

Table 9.4. *Conversation no. 4: Roger and Ally*

ADJACENT UTTERANCES[a,b]				
Novel[c]	Roger	(14)	that	(2)
	Roger tickle Ally	(9)	that that box	(1)
	Roger tickle	(3)	that shoe	(1)
	tickle Roger	(1)	Roger string George	(1)
	tickle	(1)	comb	(1)
	Roger tickle Ally hurry	(1)	good	(1)
	George	(2)	food-eat	(2)
	Joe	(7)	Ally	(2)
	string	(1)		
Expansions[d]	George smell Roger	(1)	Roger tickle Ally	(1)
	tickle hurry	(1)	Roger tickle Roger comb	(1)
	Roger tickle	(1)	Roger tickle Roger	(1)
Imitations[e]	Roger	(5)	baby	(1)
	tickle	(3)	pillow	(1)
	Roger tickle Ally	(1)	comb	(2)
	shoe tickle	(2)	pull	(2)
	shoe tickle	(1)		
NON-ADJACENT UTTERANCES[f]				
	Joe	(2)	shoe	(1)
	you tickle	(1)	pillow	(1)

[a]Numbers in parentheses indicate frequency.
[b]$N = 74$; 67 followed a teacher's command, 7 followed a teacher's question.
[c]N = 50; 43 followed a teacher's command, 7 followed a teacher's question.
[d]N = 6; all followed a teacher's command.
[e]N = 18; all followed a teacher's command.
[f]N = 5.

Table 9.5. *Conversation no. 5: Joe and Booee*

ADJACENT UTTERANCES[a,b]				
Novel[c]	food-eat Booee that	(3)	more that Booee	(1)
	food-eat Booee	(3)	that Booee	(1)
	food-eat	(3)	more Booee	(1)
	you food eat Booee that	(1)	Booee	(2)
	Booee food-eat more Booee	(1)	you Booee	(1)
	food-eat Booee hungry	(1)	more	(1)
	food-eat me fruit Booee	(1)	you	(3)
	food-eat fruit Booee	(1)	Booee hurry	(1)
	that food-eat Booee	(1)	there you gimme	(1)
	fruit food-eat you	(1)	you Booee you Booee hurry	(1)
	food-eat fruit hurry	(1)	more Booee hurry	(1)
	gimme food-eat	(1)	tickle Booee	(3)
	gimme	(2)	tickle you me	(1)
	fruit more Booee	(1)	you there	(1)
	gimme fruit hurry	(1)	hurry tickle Booee hurry	(1)
	there more fruit	(1)	you there that there	(1)
	hurry	(4)	that Booee there Booee	(1)
	fruit	(1)	baby that	(1)
	that Booee over there	(1)	more baby	(1)
	that	(4)		
Expansion[d]	hurry tickle Booee hurry Booee hurry	(1)	me tickle hurry	(1)
	Booee you Booee hurry gimme	(1)	food eat Booee that	(1)
	tickle Booee gimme	(1)	that Booee	(1)
	me tickle	(6)	that tickle	(1)
			there you	(1)
Imitation[e]	that	(4)	tickle	(1)
	you	(2)	baby	(1)
NON-ADJACENT UTTERANCES[f]				
	tickle Booee	(6)	more fruit there	(1)
	tickle	(2)	that	(1)
	more Booee	(1)	that that Booee gimme	(1)
	hurry that tickle	(1)	gimme	(2)
	tickle gimme	(1)	Booee	(1)
	food-eat	(1)	hurry gimme hurry	(1)
	food-eat Booee	(1)	that there	(1)
	that that there	(1)	that more baby	(1)

[a]Numbers in parentheses indicate frequency.
[b]$N = 79$; 32 followed a teacher's command, 47 followed a teacher's question.
[c]$N = 57$; 22 followed a teacher's command, 35 followed a teacher's question.
[d]$N = 14$; 4 followed a teacher's command, 10 followed a teacher's question.
[e]$N = 8$; 6 followed a teacher's command, 2 followed a teacher's question.
[f]$N = 23$.

signing. Little, if any, evidence is available that an ape signs in order to exchange information with its trainer, as opposed to simply demanding some object or activity.

In a typical exchange the teacher first tries to interest the ape in some object or activity such as looking at a picture book, drawing, or playing catch. Typically, the ape tries to engage in such activities without signing. The teacher then tries to initiate signing by asking questions such as *what that?*, *what you want?*, *whose book*, and *ball red or blue?* The more rapidly the ape signs, the more rapidly it can obtain what it wants. It is therefore not surprising that the ape frequently interrupts the teacher. From the ape's point of view, the teacher's signs provide an excellent model of the signs it is expected to make. By simply imitating a few of them, often in the same order used by the teacher, and by adding a few 'wild-cards' – general-purpose signs such as *give, me, Nim,* or *more* – the ape can produce utterances that appear to follow grammatical rules. What seems like conversation from a human point of view is actually an attempt to communicate a demand (in a non-conversational manner) as quickly as possible.

9.6. Future research

It might be argued that signing apes have the potential to create sentences but did not do so because of motivational rather than intellectual limitations. Perhaps Nim and Washoe would have been more motivated to communicate in sign language if they had been raised by smaller and more consistent groups of teachers, thus sparing them emotional upheavals. It is, of course, possible that a new project, administered by a permanent group of teachers who are fluent in sign language and have the skills necessary for such experiments, would prove successful in getting apes to create sentences.

It is equally important for any new project to pay greater attention to the function of the signs than to mastery of syntax. In the rush to demonstrate grammatical competence in the ape, many projects (Project Nim included) overlooked functions of individual signs (apart from their demand function). Of greater significance, from a human point of view, are the abilities to use a word simply to communicate information and to refer to things which are not present. One would like to see, for example, to what extent an ape is content to sign *flower* simply to draw the teacher's attention to a flower with no expectation that the teacher would give it a flower. In addition one would want to see whether an ape could exchange information about objects that are not in view. For example, could an ape respond, in a non-rote manner, to a question such as 'What color is the banana?' by signing *yellow* or to a question such as 'Who did you chase before?' by signing *cat*. Until it is possible to teach an ape that signs can convey information other than mere demands it is not clear why an ape would learn a grammatical rule. To put the question more simply, why should an ape be interested in learning rules about relationships between signs when it can express all it cares to express through individual signs?

The personnel of a new project would have to be on guard against the subtle and complex imitation that was demonstrated in Project Nim. In view of the discoveries about the nature of Nim's signing that were made through videotape analyses, it is essential for any new project to maintain a permanent and unedited visual record of the ape's discourse with its teachers. Indeed, the absence of such documentation would make it impossible to substantiate any claims concerning the spontaneity and novelty of an ape's signing.

Requiring proof that an ape is not just mirroring the signs of its teachers is not unreasonable. Indeed, it is essential for any researcher who seeks to determine, once and for all, whether apes can use language in a human manner. Nor is it unreasonable to expect that in any such experiment, ape 'language' must be measured against a child's sophisticated ability. That ability still stands as an important definition of the human species.

While writing *A Report to an Academy*, Kafka obviously had no way of anticipating the numerous attempts to teach real apes to talk that took place in this country and in the USSR (Kellogg & Kellogg 1933; Hayes 1951; Kellogg 1968, Gardner & Gardner 1969; Premack 1970; Fouts 1972; Rumbaugh *et al.* 1973; Temerlin 1975; Terrace *et al.* 1979). Just the same, his view that an ape will imitate for 'a way out' seems remarkably telling. If one substitutes for the phrase 'a way out' rewarding activities (such as being tickled, chased, hugged, access to a pet cat, books, drawing materials and items of food and drink), the basis of Nim's, Washoe's, Koko's and other apes' signing seems adequately explained. Much as I would have preferred otherwise, a chimpanzee's 'Report to an Academy' remains a work of fiction.

Notes

1. Portions of this chapter appeared previously in 'How Nim Chimpsky changed my mind', *Psychology Today* (1979), **13**, 65–76. The research reported in this chapter was, in part, funded by grants from the W. T. Grant Foundation, the Harry Frank Guggenheim Foundation, and NIH (RO1 MH29293).
2. Two recent reviews of *Nim* by Gardner (1981) and Gaustad (a graduate student of the Gardners 1981), have raised additional objections about the conduct of

Project Nim and my conclusions regarding the nature of ape signing. These critiques make many patently false claims: for instance, that Nim was not taught any signs during his first year and that he was only taught to sign in his 'Spartan' classroom. At best these reviews present facts stripped of their context; at worst they simply propagate misinformation. A full discussion of these issues can be found in my replies to Gardner and Gaustad (Terrace 1981*a*, *b*).

References

Bazar, J. (1980) Catching up with the ape language debate. *American Psychological Association Monitor*, **11**, 45–7.

Bellugi, U. & Klima, E. S. (1976). *The Signs of Language*. Cambridge, Mass.: Harvard University Press.

Bloom, L. M. (1970). *Language development: Form and Function in Emerging Grammars*. Cambridge, Mass.: MIT Press.

Bloom, L. M. (1973). *One Word at a Time: The Use of Single Word Utterances Before Syntax*. The Hague: Mouton.

Bloom, L. M. & Lahey, M. (1977). *Language and Development Disorders*. New York: Wiley.

Bloom, L. M., Rocissano, L. & Hood, L. (1976). Adult–child discourse: developmental interaction between information processing and linguistic knowledge. *Cognitive Psychology*, **8**, 521–2.

Bowerman, M. (1973). Structural relationships in children's utterances: Syntactic or semantic? In *Cognitive Development and Acquisition of Language*, ed. T. E. Moore. New York: Academic Press.

Brown, R. (1973). *A First Language: The Early Stage*. Cambridge, Mass.: Harvard University Press.

Chomsky, N. (1957). *Syntactic Structures*. The Hague: Mouton.

Fodor, J. A., Bever, T. G. & Garrett, M. F. (1974). *The Psychology of Language: An Introduction to Psycholinguistics and Generative Grammar*. New York: McGraw-Hill.

Fouts, R. S. (1972). Use of guidance in teaching sign language to a chimpanzee. *Journal of Comparative and Physiological Psychology*, **80**, 515–22.

Gardner, B. T. (1981). Project Nim: Who taught whom? *Contemporary Psychology*, **26**, 425–6.

Gardner, B. T. & Gardner, R. A. (1969). Teaching sign language to a chimpanzee. *Science*, **162**, 664–72.

Gardner, B. T. & Gardner, R. A. (1971). Two-way communication with an infant chimpanzee. In *Behaviour of Non-human Primates*, ed. A. M. Schrier & F. Stollinitz. New York: Academic Press.

Gardner, B. T. & Gardner, R. A. (1974*a*). Comparing the early utterances of child and chimpanzee. In *Minnesota Symposia on Child Psychology*, ed. A. Pick. Minneapolis: University of Minnesota Press.

Gardner, B. T. & Gardner, R. A. (1974*b*). Teaching sign language to a chimpanzee. VII. Use of order in sign combinations. *Bulletin of the Psychonomic Society*, **4**, 264–7.

Gaustad, G. R. (1981). Review of *Nim*. *Sign Language Studies*, **30**, 89–94.

Hayes, C. (1951). *The Ape in our House*. New York: Harper.

Hoffmeister, R. J. (1978). The development of demonstrative pronouns, locatives and personal pronouns in the acquisition of American Sign Language by deaf children of deaf parents. PhD thesis, University of Minnesota.

Hoffmeister, R. J. (ed.) (1977) *Proceedings of the First National Symposium on Sign-Language Research and Teaching*, Chicago.

Kellogg, W. N. (1968). Communication and language in the home-raised chimpanzee. *Science*, **182**, 423–7.

Kellogg, L. A. & Kellogg, W. N. (1933). *The Ape and the Child: A Study of Environmental Influence on Behavior*. New York: McGraw-Hill.

Klima, E. S. & Bellugi, U. (1972). In *Communication and Affect: A Comparative Approach*, pp. 67–00, ed. T. Alloway, L. Krames & P. Pliner. New York: Academic Press.

Kots, N. (1935). *Infant Ape and Human Child*. Moscow: Museum Darwinianum.

McIntire, M. L. (1978). *Learning to Take Your Turn in ASL*. Department of Linguistics, UCLA (working paper).

Miles, H. L. (1978). Conversations with apes: the use of sign language by two chimpanzees. PhD thesis, University of Connecticut.

Miller, G. A. (1964). The psycholinguists. *Encounter*, **23**, 29–37.

Patterson, F. G. (1978). The gestures of a gorilla: language acquisition by another pongid. *Brain and Language*, **12**, 72–97.

Patterson, F. G. (1979). Linguistic capabilities of a lowland gorilla. PhD thesis, Stanford University.

Premack, D. (1970). A functional analysis of language. *Journal of the Experimental Analysis of Behavior*, **4**, 107–25.

Rumbaugh, D. M., Gill, T. V. & Glasersfeld, E. C. von (1973). Reading and sentence completion by a chimpanzee. *Science*, **182**, 731–3.

Sanders, R. J. (1980). The influence of verbal and nonverbal context on the sign language conversations of a chimpanzee. PhD thesis, Columbia University.

Schlesinger, I. N. (1971). Production of utterances and language acquisition. In *Ontogenesis of Grammar*, ed. D. I. Slobin. New York: Academic Press.

Straub, R. O. & Terrace, H. S. (1981). Generalization of serial learning in the pigeon. *Animal Learning and Behavior*, **9**, 545–68.

Straub, R. O., Seidenberg, M. S., Bever, T. G. & Terrace, H. S. (1979). Serial learning in the pigeon. *Journal of the Experimental Analysis of Behavior*, **32**, 137–48.

Temerlin, M. K. (1975). *Lucy: Growing up Human: A Chimpanzee Daughter in a Psychotherapist's Family*. Palo Alto: Science and Behavior.

Terrace, H. S. (1979). *Nim*. New York: A. Knopf.

Terrace, H. S. (1981). Evidence for sign language in apes: what the ape signed or how well was the ape loved? *Contemporary Psychology*, **27**, 67–8.

Terrace, H. S. (1981) A Report to an Academy, 1980. *Annals of the New York Academy of Sciences*, **364**, 94–114.

Terrace, H. S., Petitto, L. A., Sanders, R. J. & Bever, T. G. (1979). Can an ape create a sentence? *Science*, **206**, 891–902.

Terrace, H. S., Petitto, L. A., Sanders, R. J. & Bever, T. G. (1980). On the grammatical capacity of apes. In *Children's Language*, ed. K. Nelson. New York: Gardner Press.

Thompson, C. R. & Church, R. M. (1980). An explanation of the language of a chimpanzee. *Science*, **208**, 313–14.

COMMENT

D. PLOOG

David Premack's book is entitled *Intelligence in Ape and Man*, not *Language in Ape and Man*. Although he was mainly interested in the two most basic language exemplars (reference relations and a sentence-generating capacity), I see his major contribution in his demonstration of the cognitive capacity of apes. This had not been dealt with since Wolfgang Köhler's book *The Mentality of Apes*. By his, as I think, ingenious methods Premack could, among many other things, demonstrate the ape's capacity to perform a feature analysis of an object, the negation of a statement, the correct use of prepositions like 'on', 'under', 'beside', and also a correct word order of up to six 'words' (*Sarah insert apple pail banana dish*). Unlike in the Gardners' training procedures, word order was taught in all the sessions. As far as I know, Premack kept statistics on word order which turned out to be significant in regard to 'syntactical' sequencing. Premack's method, as I see it, opened a new window into the mind of apes, i.e. into their cognitive capacity.

COMMENT

ROY HARRIS

Many people will find it quite unsurprising that current language-training programmes for apes have failed to produce results which convince everybody that apes can master the rudiments of human language. One reason for that failure may lie in the programmes themselves. Arguably, human infants would fare no better at the language game if subjected to comparably bizarre experiments (involving removal from natural habitat, control by members of another species, force-feeding with a relentless diet of semiotic gobbledegook, insistence on an unnatural medium of expression, etc.). It is an old principle that comparison, in order to be valid, must compare like with like. But the comparison implicit in these training programmes involves a flagrant violation of that principle.

Another reason for the failure in question may lie in the apparent confusion between two quite separate issues: (*a*) whether or not an animal is *capable* of doing *x*, and (*b*) whether or not an animal would acquire proficiency at doing *x* in the same way and via the same stages as a human would acquire a comparable proficiency. Assessments of the results of language-training programmes for apes sometimes seem to treat evidence about (*b*) as proving something about (*a*).

A third reason may lie in the fact that such assessments are based on rather nebulous views about 'the essence of human language', appealing in particular to the slippery and far from perspicuous concept of a 'rule of grammar'. Anyone who thinks that linguists are in fundamental agreement about what a rule of grammar is ought to acquaint himself at first hand with the diversity of opinion which is evident on that subject in linguistics in recent years. The range runs from those who believe that virtually every feature of linguistic structure is to be accounted for in terms of postulated 'rules of grammar' to those who hold that the notion 'rule of grammar' is itself a misconception. Even within particular schools of linguistics, quite conflicting views of what a rule of grammar is are to be found. So claims that apes have not shown any ability to operate with 'rules of grammar' are vacuous in the absence of any clear explication of what rules of grammar are.

Finally, it may be worth pointing out that anyone who thinks he does know how to define a rule of grammar in such a way that the empirical question could be resolved as to whether or not an ape is capable of mastering such rules ought to put it to the test. One's guess would be that it would very soon emerge that apes are perfectly able to handle the computational aspects of structures analogous to basic sentences and their parts. The debate would then shift back to the misguided Cartesian question as to whether the animal, although capable of doing it, 'really knew what it was doing'.

COMMENT

G. ETTLINGER

I would like to draw attention to three kinds of behaviour that I feel deserve the attention of primatologists, but have been somewhat neglected.

1. Pointing

In the human infant this skill develops by age 10–12 months, i.e. the infant can point, viz. draw the attention of its mother to an independent target, or can respond to the pointing of the mother by redirecting its gaze. Pointing is thus a simple and naturalistic act of communication. It has been described in chimpanzees, but only in situations where it might have been acquired by imitation from humans. It remains unknown whether monkeys can communicate by pointing, but M. Blaschke and I have examined this issue (to be published). Head nodding in a particular direction could be a comparable act of communication.

2. Play

Play with replicas has been described in Nim, and possibly Vicki gave evidence of symbolic play (when playing with her imaginary pull-toy).

This behaviour lends itself to a naturalistic progression from representational performance with replicas to representational performance with symbols. However, in our own unpublished observations (with E. Winner) we failed to observe either kind of play in two young chimpanzees given the opportunity to play with replicas.

3. Map-reading

Although not naturalistic, this kind of behaviour could be assessed in apes and monkeys, since they can be shown to respond to pictures, slides and films. An advantage of this complex cognitive skill is that maps with replicas of the scene, or with symbols, can be compared. We are currently assessing this skill, but it is too early to offer findings. If our monkeys could be trained to perform this task, it would add to the evidence of cognitive skills in non-human primates.

RESPONSE

H. S. TERRACE

To D. Ploog

As I have argued elsewhere (Terrace 1979), there is a marked discrepancy between Premack's claims in *Intelligence in Ape and Man* about the linguistic capacity of apes and the data he provides to support such claims. In most instances extralinguistic cues such as the homogeneity of problem-type within a particular session or the information provided by the actual referents of the so-called '6-word sentence' comprehension task (e.g. whether the pail was empty or full) provided a non-linguistic basis for solving the problem on hand. Indeed, most of Premack's endeavors in teaching 'language' to Sarah and other chimpanzees can be characterized as training to solve particular kinds of problems. In his more recent publications, Premack acknowledges that it would be erroneous to characterize the problem-solving behavior of his chimpanzees as evidence of linguistic competence (e.g. Premack 1976).

To Roy Harris

If the import of the first part of Harris's comments is that negative evidence is rarely definitive, I heartily agree. As for the difficulty of defining 'rules of grammar', it is important to distinguish that problem from the simpler problem of deciding whether *any* sort of grammar (finite-state, phrase-structure, transformational, etc.) is needed to account for an organism's combinations of symbols. The latter question can be answered by well-understood analyses which assess the extent to which one can account for the novelty and appropriateness of particular constructions by explanations more parsimonious than those which entail one or more grammatical rules.

To G. Ettlinger

It would be of interest to see the evidence of cognitive skills that Ettlinger hopes to demonstrate in the case of monkeys and chimpanzees. It should be noted, however, that none of that evidence would suffice as evidence of linguistic competence.

References

Premack, D. (1976) *Intelligence of Ape and Man*. Hillsdale, N. J.: Lawrence Erbawn Associates.

Terrace, H. S. (1979). *Nim*. New York: A. Knopf.

10. Social changes in a group of rhesus monkeys

VERNON REYNOLDS

10.1. Introduction

This chapter will be concerned with a pair of alternatives. We shall look at two alternative interpretations of a particular set of events that were observed to occur in a group of rhesus monkeys. The terms I shall be using are taken from Rom Harré's usage (this volume, see also Figure 5.1). Harré's distinction is between organic, evolved and mechanical causes on the one hand and cognitive, mental, and 'social' causes on the other. I am aware that his terms and distinctions are not without their critics. For instance, some may feel that the opposition between a causal and an intentional framework is false because intentions are susceptible to causal analysis. But, as the work of Davidson (1963) has shown, this subsumption under a common causal concept is plausible only if causes are conceived as Humean concomitances. While it is true that an intentional, like a causal, framework explains how an event is generated, this commonality misses the point. It would be extraordinary if anything were found that was not susceptible to some kind of causal analysis. But causes arise at different levels and from different sources.

The dichotomy these alternatives present is analytic. That is to say, they offer different ways of achieving an understanding of the patterns of behaviour that will shortly be described. The dichotomy is not constitutive, by which I mean it is not reflective of two fundamentally different types of underlying physical structure. As we shall see, the data can probably best be understood in terms of a conflation of the two interpretations, implying a causal structure embodying elements that function a-consciously and others that incline us to conclude that we are dealing with some kind of goal-directed, cognitive logic.

Should the two interpretations be regarded as mutually exclusive of each other? Bearing in mind that they are analytic and heuristic I think the best answer is 'yes'. In claiming that the two interpretations are exclusive of one another, it is not claimed that there are two mutually exclusive underlying causal mechanisms. A real world event when interpreted as the effect of the operation of a mechanical causal mechanism (and itself having effects via the activation of mechanical causal mechanisms), is seen as located in a certain network of relations to other such events. But when interpreted as the unfolding of a complex of meanings, leading consequentially to other semantically identified events by virtue of cognitive operations such as reference and interpretation, a real world event is seen as located in relation to other semantically identified events.

Brains are not made up of unrelated bits, but are beautifully integrated to produce co-ordinated behaviour. This co-ordination may not always appear 'logical' to us (as the example to be described will show), but it has its own logic, based on natural selection, working together with ecological circumstance. This integration of the brain does not, however, eliminate the utility of the development of mutually exclusive interpretative hypotheses where the actual mechanisms are unknown, as in the case of primate social behaviour. Their nature may be transitory: they are set up for the time being, to collapse at some future time when new insights and new terminologies enable a closer approximation of the thing studied. Thus it is in the present case: the mentation of monkeys is something which contains automatic and intentional elements fused into a unity and thus, by its nature, displays the deficiencies of the analytic starting points.

Rom Harré has in his chapter made very clear the distinction between the causal-automatic and the cognitive-intentional and there is no need to repeat this here. In his view, too, the framework in which the former is to be described is not extensible into the framework for the latter, even at the level of the stickleback. This is because identical *actions* (within the cognitive intentions framework) may consist of different *behaviours* (within the causal-automatic framework). The distinction between action and behaviour has a central place in sociological theory (Weber 1947; see also Reynolds 1980) but has not until the present (Harré, this volume) been extended into the field of animal sociology.

Following the neo-Darwinian emphasis in biology, ethology and the study of animal behaviour have kept closely to the concept of individual advantage in the explanation of animal social behaviour. Sociobiology, through the use of the concept of inclusive fitness, has been able to account for much hitherto unexplained behaviour in the social life of animals. In sharp contrast, theories involving group processes, group advantages, not to mention group consciousness, have failed to convince. The social biology of animals is a biology of individuals.

It is for this reason that the attempt to derive an alternative framework for the explanation of animal behaviour – whose roots do not lie in the (ultimately genetic) neural processes of the individual brain but in some social collective – has been unsuccessful. To the biologist, in strong contrast with the sociologist, the social collective is an outcome of the combination of selfish and 'altruistic' behaviours of individuals, evolved by inclusive fitness. There is no 'collectivity', with sanctions or any form of power to move individuals, or to which they can refer for guidance.

Thus, hitherto, animal social behaviour has had to be set apart from human social action, in which the power of the collective over the individual is fully understood. This opposition is now under critical scrutiny. Harré has shown that in the terminology used by Kummer & Bachmann (1980) there are incorporated elements of the intentionality framework, and he shows that this framework relates to our understanding and description of human social life. It does seem that monkeys, when they interact in complex social ways, are sometimes (or often) putting into practice actions resulting from a perception of the surrounding social field combined with an intentional plan for themselves in that field.

For example, Kummer, Götz & Angst (1974) have described how a male Hamadryas baboon will not approach a female if she is already with a male subordinate to himself when he first discovers her, but will sit apart and ignore her. If, however, that same male is placed with the female and there is no other male present, he will possess her. And if, then, the male who previously possessed her appears on the scene, he (the erstwhile consort) will now sit apart. What are the implications of this?

A causal-automatic interpretation of this behaviour would impute no thinking to the animals, but instead claim the existence of a simple inhibitory mechanism in the male monkey brain. This mechanism controls behaviour in the following way. There are evolved neural pathways in the brain which can be 'on' or 'off'. The major pathway controls approach to a female. If a female enters the male's perceptual field, this pathway is activated and causes him to approach her. But half-way along this neural channel another pathway feeds in. This is the avoidance pathway. Activation of this occurs if an adult male appears close to the female, in the perceptual field. This avoidance pathway overrides the approach pathway to inhibit approach, etc. The animal is a-conscious, it does not think, its behaviour is pre-determined by neural arrangements triggered by perceptual inputs from the social environment.

Such a simple model could well provide a starting point for the explanation of the observed behaviour, but as we shall see there is more to be explained than this simple model allows for. Bachmann & Kummer (1980), using an ingenious modification of the original experiments, found that the behaviour of the introduced 'rival' male was not always negative, but depended in part on the behaviour of the female herself towards her 'owner' male. Their conclusion was that the newcomer male assessed the extent of the female's preference for her owner. He avoided contact if this was strong, but if it appeared to be weak he engaged in a fight with the owner in order to gain access to the female. These additional factors led Bachmann & Kummer (1980) to the use of the intentionalist terminology that Harré has focussed on. What we must ask is: can such complex behaviour also be described in terms of the automatic-causal model?

Here two points arise. First, Harré has clarified for us the distinction between the two models. The causal model operates along given neural pathways; its operations, however complex, are pre-determined in their location and manner of function. By contrast, the intentional model lacks this rigidly determined functional system. One and the same intention may be realised in all sorts of causal ways.

The second point to note is that Kummer (1982) has questioned the validity of the 'law of parsimony' in relation to the understanding of primate social behaviour. In other words, he feels that it is debatable whether we should always prefer a simple explanation of monkey interactions over a complex one involving quite a lot of cognitive processing. His examples show that what appears a simple response, namely hiding by a female when copulating with a subadult male, must (because of the form it takes) involve a lot of unobservable spatial and social judgments.

If we now return to our earlier example, in which our intruder male appears to be taking account of the behaviour of the female and assessing the situation accordingly, we should first return to our automatic model and see if we can make it stick. In to our 'approach/inhibit-approach' neural channel we must now feed a lot more neural pathways. But we cannot specify exactly what these new channels will cause the animal to do. This is because they will not cause any behaviour at all: they are information channels, feeding data into the 'do/don't' channel.

When information shunts around a brain prior to an activity, and that activity is dependent on those shunts, then depending on what degree of shunting there is (and there could be a lot – a careful male might take a long time before finally approaching, and several billion neural messages could easily be passed in that time) the resultant activity cannot be said to be under any simple, unitary or pre-determined causal control. It looks more like what Harré calls a 'decision'. And, if the underlying mentation is 'thought', then the monkey is thinking.

For much of primate social behaviour we should thus be prepared to accept the cognitive-decision framework because we have every reason to doubt that a regular and repetitive causal chain connects a given type of stimulus situation with its response.

Harré suggests, however, that if we accept the decision framework we shall be committing ourselves to a terminology that has its roots and meanings in the description of human 'action'. Human action has its roots in the social collectivity, from whose rules action derives its structure (Harré 1979). It thus appears to be necessary, if we adopt the intentionality framework, to adopt the act-action framework as well, because the terminology will not allow otherwise.

We thus have to ask ourselves whether we want to accept the act-action framework, and the idea of a collectivity with rules guiding individual actions in the case of monkeys. In humans, such rules are easily discovered and there can be no doubt about them. But in monkeys there seems to be no evidence at all that such rules exist. Are there monkey rules that have not been described? Certainly there are cultural practices – such as the making of snowballs by Japanese macaques (Eaton 1976). But are there social rules? Chimpanzees in the Mahali mountains seem to have a convention about how grooming should be performed (McGrew & Tutin 1978). This method is not found in other chimpanzee groups. And Quiatt (this volume) makes the point that the deceptions practised by monkeys are largely 'conventionalised' rather than 'ritualised'. By this he means they are based on cognitions arising from the ongoing social process rather than being based on innate features arising from natural selection. But the comparison with humans demands something much more than this: a *bedrock* of rules concerning *all kinds* of social performances (rules of comportment, of etiquette, taboos, obligations to specified kin, role rules etc.). Such a groundswell of rules is nowhere to be found outside of human society (Reynolds 1980).

Therefore, if acceptance of the intentionality framework for primates is not divorcible from the act-action dimension of man, we should not accept it without the necessary degree of caution, or we shall find ourselves caught on the hook of terminology. What we need, in fact, is a terminology which will imply intentionality but not commit us to the act-action dimension. This would seem possible if such words as are used are given full psychological significance, i.e. are understood to imply complex cognitive thinking, but are given only a specially limited sociological significance (i.e. are understood to imply nothing about reference to a collective rule system).

Let us move on now to the second part of this chapter, which describes the events that took place in a colony of rhesus monkeys which I observed for eight months in 1960. These events provide a case study that contains elements that are germane to the ideas expressed above.

10.2. The events at Whipsnade

The Whipsnade colony which I observed consisted of some 20 animals. There were four adult males and five adult females. The adult males and the females had formed separate hierarchies. The dominant adult male was named Henry. The dominant adult female was named Anne. These two formed a consort pair and were usually found grooming each other or just sitting together. When Henry moved around, Anne usually moved around with him although occasionally the two would split up. I shall not give the names or any details of other monkeys in the group except for two other adult females who were important in the ensuing series of events. The second most dominant female was Malvolia who was not able to approach Henry because Anne would keep her away. Anne was in fact aggressive to all the other adult females but especially so towards those who tried to approach Henry. What she would do when another female came close to Henry or tried to groom him was to threaten the female concerned by threat vocalisations, staring and bobbing her head up and down, and then if this did not succeed in causing the other female to move away, she would quite frequently attack the female concerned by chasing her, grabbing her and occasionally biting. The other female who must be mentioned was Blondie. Blondie was the most submissive female in the group, who kept very much away from Henry and from Anne and indeed from any activities.

An important aspect of the relationship between Henry and Anne was that whenever Anne chased another monkey, Henry almost immediately responded by chasing and attacking Anne. Thus, whenever Anne chased away Malvolia, Henry would chase and attack Anne. The curious thing about this relationship was that while Henry would attack Anne when she was behaving aggressively, he behaved very positively towards her most of the time and his occasional onslaughts against Anne did not appear to affect the continuation of his positive grooming relationship with her, and their consortship in general. Anne nevertheless suffered considerably as a result of these attacks by Henry. Rhesus monkeys have large canine teeth, and when Henry attacked Anne he quite often drew blood. In the end, after I had been observing the group for about a month, Henry launched such a vicious attack on Anne that he tore her leg very severely. The Zoo's veterinarian, on seeing Anne, decided that she must be removed from the colony. Attempts to remove her failed, however, since she could still climb to the top of the tree in the middle of the colony and it was therefore decided that she should be shot. This was duly accomplished and her dead body was removed.

For three days after this event the group showed great signs of disturbance, but then things began to settle down. The first clear new development was that Malvolia began to follow Henry and to attempt to groom him. Henry, for his part, did not appear to welcome the attentions of Malvolia. He did not accept her presence near him but frequently moved away when she came and sat beside him. He did not respond to her grooming by grooming her, but ignored her attentions and occasionally turned on her and chased her away. These events became more frequent with Henry occasionally attacking and biting Malvolia. Nevertheless, his attacks on her were quite ineffective in getting her to stop following him around. Her appearance changed considerably: she got thinner, her hair became ruffled as a result of constantly being chased and attacked, and her eyes began to narrow as if she were in pain. She nevertheless continued to follow Henry around and try to interact with him. This process continued for over a month.

Meanwhile, Henry began to develop an interest in Blondie. At first Blondie did not respond positively to Henry's attention in any way. She kept away from him if she could, she bared her teeth when he approached her and seemed generally to be uneasy in his presence. Henry nevertheless persisted in following Blondie and grooming her. After a week or two, Blondie became more used to Henry's attentions and allowed him to groom her and sit with her and did not show the fear grimace that had been previously seen.

Malvolia during this period not only persisted in following Henry, but became hostile towards Blondie. She chased Blondie, and occasionally attacked her without doing her any serious harm because her attacks were not very well co-ordinated, and she seemed afraid of antagonising Henry during attacks. Blondie, having become used to Henry's attentions, began slowly but surely to become more resistant to Malvolia's attacks. Eventually Blondie jumped from a branch of the tree onto Malvolia (sitting below the tree), and bit her. Malvolia at this point ran away from Blondie and from that moment the dominance relationship between the two females was reversed. Not only this, but Blondie appeared to be fully established as Henry's consort. She adopted a behaviour pattern similar to that Anne had followed previously, by chasing away other females who came close to Henry. Henry continued a close relationships with Blondie, but during the course of the last few months of my observations, I noticed that as Blondie began chasing away other females, Henry would himself launch attacks on Blondie, just as he had previously done with Anne.

10.3. Interpretation

That account more or less sums up the series of events that took place at Whipsnade during 1960 (see Reynolds 1962). There seem to be two aspects of the behavioural sequences described which indicate that automatic processes were taking place during the period described. In the first place there was Henry's reaction to aggressive behaviour by his consorts Anne and Blondie. I was struck at the time of observation by the fact that Henry appeared quite incapable of resisting the urge to attack Anne (or Blondie) when she was behaving aggressively. He did not show any compensation for the fact that the animal he was attacking was his consort. Being dominant male,it was frequently the case that he attacked animals who behaved aggressively: usually these were sub-dominant males. This behaviour was readily understandable in terms of his position as the dominant male. We could say, for instance, that evolution has favoured responses in the adult male that cause him to behave aggressively to other animals offering challenges to his dominant position. It could be argued that it was this aggression towards aggressive behaviour in other animals that was the characteristic of the dominant male, and that it was the animal which was best able to execute this innate strategy that was able to become and maintain himself as the dominant male. Why, however, attack his consort? The only answer I am able to suggest is that a dominant male either cannot or does not distinguish between his consort and a rival. He appears to be responding not to the particular animal emitting the signal of aggressive behaviour but to that signal itself. In other words, it is the behaviour of another animal – rather than the animal itself behaving in a certain way – that triggers his aggressive response.

This suggestion is in direct opposition to the view expressed by many primate ethologists, i.e. that the response of a monkey to a situation in a social group is not specifically to the behaviour being exhibited by other animals but very much to the specific animal giving that behaviour. Thus an action by another monkey may not be responded to, and it is often held to be typical of monkeys to take the individual characteristics of the animal giving the signal into account before reacting. This did not seem to be the case with Henry.[1]

A second feature which seemed to be of a rather automatic kind was the persistence of Malvolia's following of Henry. Here again we could understand this as a case of a piece of programmed behaviour of the following kind. A female is programmed to go near to and eventually to present herself sexually to a dominant adult male if no other adult female has already established herself in this way. Females thus in competition over access to a dominant male establish a hierarchy, and the animal who is successful in establishing herself at the top of this hierarchy is the one who consorts with the dominant male, and stands the greatest chance of successful reproduction. The presence of a female with a dominant male acts as a signal inhibiting the association of another female with this male. Thus while Anne was with Henry, and in view of Anne's success over Malvolia in aggressive interactions between them, Malvolia did not succeed in usurping Anne's position, but had to content herself with second place. After Anne's disappearance, Malvolia found the central obstacle to her associating with Henry had gone and therefore she could move in. However, she persisted in following Henry for so long and despite so much aggression from him, that her continual following of him appeared to be quite unlike the behaviour of an animal which was taking into account the disadvantages of this kind of behaviour. She appeared not to be counting the cost of her attempts to associate with Henry. It was as if she was completely incapable of realising that Henry was not going to allow her to consort with him. Thus the second aspect of the behaviour of these animals that can be taken as indicating an almost automatic process was this persistence of Malvolia in the face of constant punishment.

So far the two processes I have considered have related to a rather complex series of behavioural interaction. This automatic dimension can, however, be pursued far more thoroughly down to the level of individual units of behaviour (see Reynolds 1970). In such a system, the emission of a particular behavioural signal by one animal is responded to by a particular behavioural unit emitted by another animal, which in turn acts as a stimulus to the behaviour of the first animal, etc. This series of interactions can be said to be the outcome of natural selection and to constitute the fabric of social interactions in a group. It is the result of innate processes and serves to further the interests, and in particular the reproductive success, of each individual, The outcome of the entire process of individual actions and reactions, knitted together in these innate ways, is the social fabric of the group itself. This kind of interpretation of behaviour is very much in keeping with the understanding of the social interactions of animals throughout ethology. In particular, the interactions of species low in the phylogenetic scale (for example, insects, fishes, and to some extent, birds), can successfully be understood in this way.

We know too that behaviours not overtly displayed are latent, for a change in the environmental or social circumstances can dramatically alter an animal's behaviour: the dominant becomes subordinate and vice versa. These changes, in the automatic model, will be seen as the outcome of neural processes perhaps resulting from changes in hormone levels or other physiological processes.

The fact that these automatic changes occur is not in any way haphazard: their structure has been shaped by natural selection. As a result we can refer to these changes as 'adaptive', and to a whole pattern of behavioural changes, taken together, as a 'strategy'. The term 'strategy' has the great advantage that it can be used in connection with the causal-automatic framework, as well as the cognitive-intentional one.

A second method of understanding the behaviour of the Whipsnade rhesus monkeys relies freely on the attribution of thinking – the ability of the monkeys themselves to make cognitive decisions and to plan strategies. This approach makes the described interactions much more readily comprehensible in human terms, and seems to demand far less of the imagination than the objective description given above. All we need do, if we follow this method, is to say that Henry was an aggressive, dominant male who did not like his consort behaving aggressively to other animals, and therefore wanted to punish her when she did so. Malvolia was keen to become the consort and, when the opportunity arose with Anne's death, she made great efforts to gain Henry's company. Henry, however, did not like Malvolia and attempted to get her to stop her attentions by attacking her. Henry was attracted by Blondie and wished to make her his new consort; Blondie was afraid of Henry and so kept out of his way. Malvolia, seeing Henry's attentions to Blondie, attacked Blondie in an attempt to intimidate her, but Blondie, because of Henry's attentions, became more dominant and eventually subdued Malvolia and became Henry's consort. Again, however, since Henry did not like aggressive females, he attacked Blondie when she behaved aggressively. This description attributes the ability to work out quite complex moves and strategies in the achievement of particular objectives, and relies heavily on a humanistic interpretation of the monkeys' behaviour, or at any rate on an anthropomorphic conception of the nature of the cognitive machinations of monkeys.

The question now is: to what extent should we accept the automatic-causal description, and to what extent ought we to interpret the monkeys' behaviour in terms of plans? As we have seen, Rom Harré regards the interpretation in terms of the automatic-causal framework as irreconcilable with the interpretation in terms of the intentionality framework. In fact, Harré's contrast is not so much between the actual processes of behaviour and action, as between the two conceptual systems in which the two levels of phenomena operate.

Looking at the analysis of primate behaviour by Kummer & Bachmann (1980), he regards the explanation they give as 'transitional' because it uses terms derived from both the automatic framework and the intentionality framework without distinguishing the two clearly. In this sense the description by Kummer & Bachmann is indeed transitional between two conceptual systems, but what if the monkeys themselves are transitional between two forms of communication? In that case the transitional frame of reference used for conceptual systems spotted by Harré is the one most likely to succeed in giving us an understanding of primate behaviour. It may thus be that even though, as Harris says (this volume), some kind of thought is always present in every species that has any kind of communication process, being simply a 'logomachy' deriving from the ongoing process, there may nevertheless be a value in making a distinction between earlier (in evolutionary terms) systems of communication based largely on automatic processes of observable signal-response chains, and (evolutionarily) later communication systems in which the presence of unobservable cognitive components becomes progressively more characteristic of the communication process. This is most evident in humans, but is to some extent evident in other primates.

The advantage of a more cognitive neural process underlying behavioural outcomes is that, to the extent that neural planning could take account of the immediate parameters of social situations, the final behavioural outcome or set of actions would be likely to be more appropriate to the situation concerned. Thus we can imagine that in the evolution of primates, the complex information processing of the monkey brain already has gone some way towards cognitive planning of strategies but has not so far reached the degree of sophistication found in the highest primates (chimpanzees or humans).

As stated earlier in this chapter, even though two types of interpretation may be mutually exclusive analytically, there is no reason to suppose that both might not be needed in the explanation of primate social behaviour, and indeed it seems that both are needed. In such an explanation, the necessary first step would be to try and identify those aspects of the behaviours observed that appear to result from the working out of evolved and relatively inflexible neural cause-and-effect processes, as against those which show a flexibility in which the outcome is contingent on what appears to be a cognitive assessment of relevant environmental criteria.

Whether we use my original interpretation of Henry's attacks on aggressive females, or Seyfarth's suggestion (see Comment), we are left with a process at the heart of rhesus social life that is rather automatic. It is an evolutionary product, an innate neural response. Whether the selection pressure behind it is male–male rivalry or male–female peace-keeping, the process described does not seem a likely candidate for the intentionality framework. If Henry had been 'thinking' before his attacks on Anne, it is likely either that he would have 'decided' not to attack at all, or that he would have moderated his bites. This would be 'logical', in view of the fact that Anne was, for 99 percent of the time, his closest associate and grooming partner. Furthermore, we have the persistence of Malvolia's approaches to Henry. Here again, the intentionality framework seems to hit problems. Perhaps Malvolia thought about the situation after Anne's death and decided to make a bid for Henry. She had, after all, been the second, and was now the top ranking female. Such a move was eminently 'logical'. Let us even assume that her initial moves towards Henry were prompted by a cognitive plan. Would not a cognitively motivated monkey, thinking about what it was doing, have given up as day after day and week after week went by with nothing but negative reinforcement? Would not the 'message', so painfully obvious to myself as observer, have got across? Might we not expect that even a relatively small amount of discouragement from Henry would cause her to desist? How can we explain her protracted persistence, while she became progressively more and more pathetic as Henry bit her more often, and her rival Blondie grew more confident?

Perhaps some in reading this will feel that, curious though these monkey persistencies are, they are certainly also found in humans. If so, and if they are largely evolved, automatic responses, then to that extent we need to include psychological automatisms in humans.

So, to draw to a conclusion, I feel we are in very great difficulty in the description of primate social behaviour. On the one hand, ethology provides us with a form of description and elements of a terminology that are most useful in the description of species-specific, limited, rather precisely channelled behaviour patterns, from which relationships and social patterns can be derived. On the other hand we have at our disposal a range of terms developed over the centuries for the description of human action in all its complexity. Monkeys and apes are now forcing us to look again at the standard ethological interpretations characteristic of the last couple of decades. Workers such as Bachmann & Kummer (1980) have shown the extent of subtlety underlying primate social performances. In the rhesus monkeys of Whipsnade this was clearly evident in the everyday happenings of the group. Often I was able to predict whether a monkey would move or not, or to which other monkey it was walking, on the basis of my knowledge of their past interactions. They planned their activities most of the time. But along with the surface decision-structure there was a deeper core of innate regularity. The automatic and the intentionality frameworks will both prove useful, in future years, in elucidating the components underlying the social behaviour of monkeys and apes, and students of this topic will have to be ever more careful to note the exact meaning of the words used in their descriptions.

Notes

1. I have let my original text stand. In the meantime, Seyfarth in a comment at the end of this chapter has offered a potentially more convincing explanation of Henry's behaviour. We should note, however, that the new explanation remains firmly in the 'automatic-causal' paradigm.

References

Bachmann, C. & Kummer, H. (1980). Male assessment of female choice in Hamadryas baboons. *Behavioral Ecology and Sociobiology*, **6**, 315–21.
Davidson, D. (1963). Actions, reasons and causes. *Journal of Philosophy*, **60**, 685–700.
Eaton, G. G. (1976). The social order of Japanese macaques. *Scientific American*, **235**, 97–106.
Harré, R. (1979). *Social Being*. Oxford: Blackwell.
Kummer, H. (1982). Social knowledge in free-ranging primates. In *Animal Mind – Human Mind*, ed. D. R. Griffin, pp. 113–130. Berlin: Springer.
Kummer, H., Götz, W. & Angst, W. (1974). Triadic differentiation: an inhibitory process protecting pair bonds in baboons. *Behaviour*, **49**, 62–87.
McGrew, W. C. & Tutin, C. E. G. (1978). Evidence for a social custom in wild chimpanzees? *Man*, **13**, 234–51.
Reynolds, V. (1962). The social life of a colony of rhesus monkeys (*Macaca mulatta*). Ph.D. thesis, London University.
Reynolds, V. (1970). Roles and role change in monkey society: the consort relationship of rhesus monkeys. *Man*, **5**, 449–65.
Reynolds, V. (1980). *The Biology of Human Action*. Reading: Freeman.
Weber, M. (1947). *The Theory of Social and Economic Organization*. London: Free Press.

COMMENT

ROBERT M. SEYFARTH

Reynolds describes an intriguing series of events in his colony of captive rhesus monkeys, and mentions his puzzlement at the fact that his alpha male, Henry, would chase alpha female Anne every time she attacked another female. Henry's attacks occurred despite an otherwise amicable relationship with Anne, and ironically led indirectly to her death.

The behavior described by Reynolds is strikingly similar to that reported by Downhower & Armitage (1971) for the yellow-bellied marmot. In marmots there is an interesting 'conflict between the sexes' over the group size that leads to the optimum number of offspring. On the one hand, male reproductive success is a direct function of harem size. A male will leave an increasing number of offspring as the size of his harem increases. On the other hand, female reproductive success *decreases* with increasing harem size. A female, on average, will leave the most offspring if she is the only female, fewer offspring if there are two females in the harem, and fewer still if she is part of a harem of three females.

The conflicting demands of males and females become most apparent when there are one or more new females trying to enter an already established group. It is in the male's reproductive interest to establish increasingly large groups, while it is in the female's reproductive interest to keep the newcomers out. Assuming that males and females will act to maximize reproductive success, we can predict both aggression among females and attempts by the male to keep this aggression to a minimum.

Sociobiological theory offers a plausible explanation for the observations described in Reynolds' paper. Such theory does not, however, allow us to distinguish between automatic responses and cognitive strategies in the animals involved.

Reference

Downhower, J. F. & Armitage, K. B. (1971). The yellow-bellied marmot and the evolution of polygamy. *American Naturalist*, **105**, 355–70.

COMMENT

DUANE QUIATT

Reynolds outlined two alternative interpretations of behaviour, 'automatic processes' v. 'cognitive intentions'. He observed that not all behavior signals in a species' (or individual animal's) repertoire may be available to an individual at any given time, and he noted the utility of a concept such as *strategy* for discussing the regulation of behavior. We can talk about strategies and objectives, he said, without specifying whether or not an animal has conscious intentions regarding the strategy it is enacting.

Implicit in his discussion was the idea that a strategy enacted more than once might employ different signals depending on the availability of signals to the actor. Would the question: 'in which conditions might alternative signals be used in enacting the same strategy?' lead us back to the issue of 'automatic processes' v. 'cognitive intentions'? Would it be useful to adopt an intermediate terminology (e.g. 'control features'), thus focusing attention on the need to specify under what general conditions identical strategies can be pursued, in what interactional contexts, and with what likelihood of success? Some such delineation of the ecology of behavioural action seems to me to be in accord with Reynolds' suggestion.

COMMENT

H. S. TERRACE

Most psychologists and philosophers still accept Descartes' division of organisms into two groups, human (with the capacity to think) and animal (without it). According to this point of view, animal behavior, no matter how elaborate and complex, can always be reduced to some configuration of reflexes in which thought plays no role. It is, of course, true that this attitude presupposes a number of important extensions of Descartes' model of the reflex, in particular, different types of conditioned reflex (cf. Skinner 1938). It is, however, equally true that the different models of the conditioned response that have been proposed by twentieth century behaviorists have been regarded as automatic and thoughtless reactions to some immediate stimulus.

The success of stimulus–response (S–R) models of conditioned behavior has led most modern learning theorists to argue that nothing is gained by trying to explain an animal's behavior by appeal to its mental life and that the sole mission of a behaviorist is to isolate and to manipulate environmental variables which control behavior (e.g. Skinner 1950; 1974). Appeal to unobservable mental events (or to neural events or to activities of a conceptual nervous system), at best detract from the proper concern of a behaviorist: prediction of behavior as a function of environmental variables.

Alternatives to the S–R model have hardly gone unnoticed. Almost 70 years ago, Hunter succinctly characterized the nature of evidence that would allow one to argue that an S–R model was not sufficient to explain some instances of animal behavior. 'If comparative psychology is to postulate a representative fact, . . . it is necessary that the stimulus represented be absent at the moment of response. If it is not absent, the reaction may be stated in sensory-motor terms' (Hunter 1913:21). The few attempts (e.g. Tolman) to develop alternatives to the S–R approach have met with little success. Purported instances of animal cognition were effectively neutralized by S–R theorists who postulated various types of covert mediators that seemed to satisfy Hunter's 'sensory-motor' type of explanation. As a result, behaviorists had little reason to concern themselves with what went on inside an animal's head.

This state of affairs has changed markedly during the past five years. A variety of studies on spatial memory (Olton 1978; 1979), on delayed matching to sample (Roitblat 1980), on sensitivity to monotonic patterns of food magnitudes (Hulse 1978), on memory of the locus of key pecks in a sequence of pecks to different keys (Shimp 1976; Shimp & Moffitt 1974), on recall of photographs (Wright *et al.* 1983), on recognition of sequences of different stimuli (Weisman *et al.* 1980; Weisman & DiFranco 1981), and on serial learning (Straub & Terrace 1981, Terrace, 1983); have created unprecedented difficulties for an S–R model of learned behavior. In each of these studies the behavior in question could not be explained by reference to an immediately available stimulus or to some covert S–R mediator. Accordingly, it appears as though Hunter's conditions for postulating a 'representative fact' have finally been met and that, for the first time, psychologists can argue that animals have the capacity to think.

The existence of stimulus–stimulus (S–S) representations in animals poses a fascinating problem in that it provides a basis for studying complex forms of memory in non-verbal creatures. Theories of human memory (e.g. Anderson & Bower 1973) invariably postulate some sort of verbal or symbolic mediation. Accordingly, the ability of an organism to solve problems and to order events in its memory without verbal mediation should have two important consequences. It should provide an invaluable biological benchmark for comparing human and animal memory. It should also provide psychologists with the opportunity to study the intriguing problem of how a creature thinks without language.

References

Anderson, J. R. & Bower, G. H. (1973). *Human Associative Memory*. Washington, DC: W. H. Winston & Sons Inc.

Hulse, S. H. (1978). Cognitive structure and serial pattern learning by animals. In *Cognitive Processes in Animal Behavior*, ed. S. H. Hulse, H. Fowler & W. K. Honig, pp. 311–40. Hillsdale, NJ: Lawrence Erlbaum Associates.

Hunter W. S. (1913). The delayed reaction in animals and children. *Animal Behavior Monographs*, **2** (1).

Olton, D. S. (1978). Characteristics of spatial memory. In *Cognitive Processes in Animal Behavior*, ed. S. H. Hulse, H. Fowler & W. K. Honig. pp. 341–74. Hillsdale, NJ: Lawrence Erlbaum Associates.

Olton, D. S. (1979). Mazes, maps and memory. *American Psychologist*, **34**, 588–96.

Roitblat, H. L. (1980). Codes and coding processes in pigeon short-term memory. *Animal Learning and Behavior*, **8**, 341–51.

Shimp, C. P. (1976). Short-term memory in the pigeon: relative recency. *Journal of the Experimental Analysis of Behavior*, **25**, 55–61.

Shimp, C. P. & Moffitt, M. (1974). Short-term memory in the pigeon: stimulus response associations. *Journal of the Experimental Analysis of Behavior*, **22**, 507–12.

Skinner, B. F. (1938). *The Behavior of Organisms*. New York: Appleton-Century-Croft.

Skinner, B. F. (1950). Are theories of learning necessary? *Psychological Review*, **57**, 193–216.

Skinner, B. F. (1974). *About Behaviorism*. New York: Alfred A. Knopf.

Straub, R. O. & Terrace, H. S. (1981). Generalization of serial leaning in the pigeon. *Animal Learning and Behavior*, **9**, 454–68.

Terrace, H. S. (1983). Simultaneous chaining: the problem it poses for traditional chaining theory. In *Harvard Symposium in the Quantitative Analysis of Behavior*, ed. M. L. Commons & A. R. Wagner. Cambridge, Mass.: Ballinger.

Tolman, E. C. (1932). *Purposive Behavior in Animals and Man*. New York: Appleton-Century-Croft.

Weisman, R. G. & DiFranco, M. P. (1981). Testing models of delayed sequence discrimination in pigeons: Delay intervals and stimulus durations. *Journal of Experimental Psychology: Animal Behavior Processes*, **7**, 413–24.

Weisman, R. G., Wasserman, E. A., Dodd, P. W. & Larew, M. B. (1960). Representation and retention of two-event sequences in pigeons. *Journal of Experimental Psychology: Animal Behavior Processes*, **6**, 312–25.

Wright, A. A., Santiago, H. C., Urcuioli, P. J. & Sands, S. F. (1983). Monkey and pigeon acquisition of same/different concept using pictorial stimuli. In *Harvard Symposium in the Quantitative Analysis of Behavior*, ed. M. L. Commons & A. R. Wagner. Cambridge, Mass.: Ballinger.

REPLY

V. REYNOLDS

I am grateful to Seyfarth for his very interesting suggestion as to the explanation of Henry's behaviour. If correct, this must imply an element of genetically mediated causation, perhaps in the form of an underlying neural sex difference. From such a difference, cognitive strategies appropriate to the immediate circumstances could arise during adult life. Quiatt's point that there should be a closer study of the use of alternative signals in pursuance of the same strategy seems a very valuable one, and would certainly help to elucidate *what* the underlying strategy was, and *how* a particular individual chose to carry it out. This too might shed light on the processes involved in animal thinking referred to by Terrace, who rightly directs our attention to relevant psychological work relating to this issue.

B. Internal and external environments of social signals

11. Categorization of social signals as derived from quantitative analyses of communication processes

M. MAURUS AND D. PLOOG

11.1. Introduction

One aspect of the problem with which this volume is concerned is the relation between meaning and function of signals. As it has been stated by Reynolds and Harré (this volume), the function of signals is usually thought of in terms of the discernible effects of a signal on the recipient. The meaning, on the other hand, refers to the unobservable intention in the brain/mind of the animal giving the signal and the corresponding interpretation by the receiver.

This chapter deals entirely with the function of signals and cannot make any claims for explaining the meaning. However, the categorization of social signals, as attempted here, may provide useful boundary conditions for discussion of the meaning of signals, i.e. on the unobservable intention in the mind of the primate and the corresponding interpretation by the receiver.

As will be shown, the analysis of signal function poses complex problems which can be overcome only in part. Provided the function of signals is usually thought of in terms of the discernible effects of a signal on the recipient, our question is: What part of the effect of a signal is discernible? And what part of that which can be actually seen is the effect of the signal? It may be assumed that, in primates, the behaviour of a recipient is not only influenced by a signal from another animal but also by other external factors and by the motivational state of the animal. In other words, the behaviour displayed by an animal after receiving a signal from another animal is not necessarily affected only by the signal received. The following transitions, from a preceding behavioural event shown by the sender to an immediately following behavioural event by a conspecific, are possible and likely to be regarded by an observer as a sequence of signal and effect on the recipient:

(*a*) The conspecific notices the signal and reacts discernibly.
(*b*) The conspecific notices the signal and reacts indiscernibly to the observer.
(*c*) The conspecific notices the signal but does not react to it.
(*d*) The conspecific notices the signal, does not react, but shows behaviour caused by influences other than the signal received.
(*e*) The conspecific notices the signal, reacts to it and to other, e.g. previous, influences.
(*f*) The conspecific does not notice the signal but reacts to other influences.
(*g*) The conspecific does not notice the signal and does not react.

To an observer, it will be extremely difficult to distinguish the sequences (*b*) – (*g*) from sequence (*a*) unless he is familiar with the function of the communicative signals.

Nevertheless, sequence (*a*) does occur frequently enough to stand out clearly against the other variations whenever the signal is given often enough. It will be observed more often than any of the other combinations for the following reasons. Provided the observable behavioural event by the sender is a signal with a discernible effect on the recipient, the recipient will – at least occasionally – react to it. If he does not react or reacts wrongly (from the sender's point of view), the sender will repeat the signal or send others in order to achieve the corresponding effect on the recipient (Kummer, Goetz & Angst 1970). Statistically, this will place the highest frequency rate of the transitions from the signal to the discernible effect with the recipient.

There are cases, however, where a signal is repeated frequently under stable conditions (same sender, same recipient, same test conditions) and still cannot be assigned with statistical probability to specific reaction patterns of the recipient. This may be due to the following reasons. Of the behavioural repertoire of a species, the observer registers certain units which he – in the absence of objective criteria and more or less at his own discretion – designates as signals. Proof for the reliability of such a clear assignment of units to the specific types of signals used by the animals, i.e. proof that each unit corresponds to a specific type of signal (function) and each type of signal to a unit, is still missing. In support of this statement, reference is made to the studies by Reynolds (1976) in which several authors were cited, each of them offering a different categorization of the behavioural units for one and the same primate species.

The goal of the methods we developed was the categorization of the squirrel monkey's repertoire of visually recognizable signals in the area of agonistic interaction. In these experiments, our primary objective was to find out which of the units selected by the observer had the same function and which of them differed in function, rather than to find out which functions the units had. For tackling this problem, a basic requirement was that none of the units selected by the observer had more than one function. The function, however, is not known prior to the analyses, so attempts at defining it could be made only by intuition.

The analyses are based on a compilation of units with the same function. Errors in splitting the repertoire, i.e. making the units too small, will not adversely affect the final results, but they will certainly cause more work in collecting data and evaluating them. In the case of units being too large, i.e. having more than one function for the animals, the analyses will indicate the need for division of the units. This procedure may be repeated with arbitrarily selected units until correct results are achieved.

Below are some examples of the methods we used in categorizing the behavioural repertoire of the squirrel monkey by functional criteria. It is intended to give a description of the methods applied and objectively demonstrate their reliability, rather than exhaustively cover and interpret the social behaviour of the species.

The first step, the classification of units on the basis of similar or differing functions, was performed exclusively by applying methods used in quantitative measures, statistics and information theory. The second step, categorizing the classes found (by 'categorizing' we mean the identification and labeling of types of function), was derived from the differences between classes. A class is equivalent to the set of all units having the same function.

11.2. Classification of units

Experimental procedure

When investigating the relations between a signal and its discernible effects, it is advantageous to create a situation which permits the sender the frequent repetition of his signal. We have succeeded in doing this in our experiments (Maurus 1967; Maurus & Ploog 1971; Maurus & Pruscha 1972). Out of a group of squirrel monkeys one was selected and, by remote-controlled electrical brain stimulation made to show a certain reaction – a procedure which can be repeated at brief intervals. This forced reaction served as a signal which affected the recipients discernibly. Numerous communication processes followed, all of them starting with the same behavioural event. They furnished the data needed for quantitative analyses. In this connection, the electrical brain

Table 11.1. *List of behavioural units and classes*

Unit description	Unit no.	Class no.	Category label
Touching partner's head	05	91	Threat, low intensity
Touching partner's hip	06		
Touching partner's back	07		
Touching partner's tail	09		
Touching partner's extremities	11		
Advancing mouth towards partner's neck	12		
Mounting	21		
Straightening body in front of partner	14	92	Threat, medium intensity
Thrusting chin towards partner	15		
Jumping onto partner	23	95	Threat, maximum intensity
Beating partner	24		
Biting partner	25		
Genital display	18	93	Dominance gesture
Inspection of partner's genitals	19		
Advancing nose towards partner's sitting place	31		
Lolling, sprawling	08	94	Triumph gesture
Back-rolling with presentation of ventral view	20		
Running away	27	97	Submissive gesture

stimulation served merely as a tool for evoking communication processes at the time desired. It had no falsifying effect on the results of the categorization procedure (Maurus *et al.* 1981).

The communication processes were filmed and evaluated frame by frame. Any visually recognizable behavioural Unit (see Table 11.1, columns 1 and 2) of the sender and the recipient and the time of occurrence was coded and punched on paper tape for subsequent computer analysis. The behaviour which was elicited by brain stimulation to start the interaction sequence was not included in any of the analyses as it might be considered an artifact.

Evaluation procedure

(a) Transition frequencies

With certain restrictions, any behavioural event in a communication process may be considered a signal and any subsequent behaviour by a conspecific as a discernible effect. Under this assumption, several transitions from the preceding to the succeeding behavioural event are observable per communication process. Each of these transitions may be regarded as a sequence of signal and discernible effect. If the description of communication processes meets the demand that each unit have no more than one function, units with the same function can be recognized by the fact that they induce the same reaction. Here is an example: if unit A is followed by C in 10 percent and by B in 90 percent of cases, and if F is followed by C in 10 percent and by B in 90 percent of cases, A may be assumed to have the same function as unit F. On this basis, it will be possible to count the transition frequencies between units and find out where these transition frequencies show similarities. A transition matrix is shown in Figure 11.1, with the rows depicting the preceding and the columns the succeeding units. This figure clearly demonstrates that it is possible to combine the individual units into groups (see Table 11.1, columns 1, 2 and 3). Within each group, there is a great similarity in transition frequencies. The differences between the various groups are considerably greater than those within each group. This relatively simple method (Maurus *et al.* 1979) opens an initial approach to a classification under functional aspects.

(b) Cluster analysis

Another, more sophisticated method – also based on the transitions from the preceding to the succeeding behavioural event – is cluster analysis, which can also furnish data on the similarity of units. In the first step of this method, from the set of all units (n) the two with the greatest similarity are selected and combined in one unit. The next step is to pick out from the new set of units ($n-1$) again those two with the greatest similarity and combine them. This procedure is continued until the optimum number of clusters is found. By cluster analysis, the same type of classification is achieved as by the simpler method of transition frequencies described above (cf. Maurus & Pruscha 1973).

Considering the transitions from a preceding to an immediately following behavioural event is not the only starting point for classifying the units under functional criteria. There is little doubt that prior events, such as the penultimate one or the one before that, also influence the behaviour of a conspecific. This influence does not only depend on how far back this event occurred, but also on how long its effect on the continuing communication process has lasted. With the multi-response linear learning model the intensity of this effect can be calculated (Pruscha & Maurus 1979). The calculation of the intensity values has

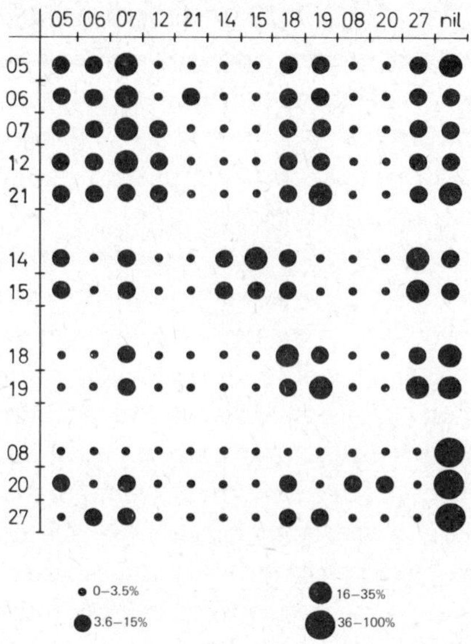

Figure 11.1. Frequencies of transitions from preceding behavioural unit (row) to immediately succeeding behavioural unit (column). Each row adds up to 100 percent.

verified the classification so far achieved by the other methods described. Not only do the intensity values provide indications for classifying the units, but they also give indications for assessing the function of the classes established (see below).

(c) *Position of units*

Another indicator for categorization is the position of the units within the communication process. Very favourable conditions for assessing such processes are established by electrical brain stimulation: the event induced by it causes a sequence of rapidly succeeding events which lasts from several seconds to a maximum of one minute. After a reasonable interval (e.g. 20 min), the next communication process may be induced. This enables the experimenter to separate one communication process clearly from the next one and to determine an observable beginning and end of each process. As all communication processes start with the same event, they provide a good data base for statistical analyses. In as much as, within a communication process, the events succeed each other rapidly, the duration of the process can be represented by the number of events observed without considering the variations of the intervals between the events.

The experimenter may then want to examine whether certain units occur in specific positions (e.g. the ultimate or penultimate one) of a communication process more often than others, or whether they occur in specific positions particularly seldom. When assuming that a sequence of signals with a specific function does not occur at random but is subject to certain rules, then those units that assume specific positions within a process may safely be assigned to specific functions. Other units assuming that position can then be assigned the same function. As not all of the communication processes have the same duration, they cannot simply be added up. Only sections of them – the beginning, the middle or the end – may be considered without risking major errors attributable to the difference in the duration of the processes. Figure 11.2 shows, for example, the frequencies with which the units occur in the final four positions of a communication process. The frequencies registered in the last two positions, t and $t-1$, or, more specifically, the differences between them, are the most conclusive. It is evident that this method produces the same classes as the ones shown in Figure 11.1.

From the unit positions in a communication process a number of other calculations may be derived. When the total duration of a process is considered, for instance, cluster analysis may be applied again. This analysis is similar to the method applied to the transition frequencies between preceding and succeeding behavioural events. The only difference is that cluster analysis in this case is based on linking the events with their positions in the communication process. This method verifies once more the classification described earlier (Maurus & Pruscha 1973).

(d) Variation of experimental conditions

Other classification methods consider the frequency changes observed in the occurrence of the units under varying experimental conditions (e.g. change of intervals between stimulations, or in group composition). Such methods are based on the assumption that units that change their frequency under varying conditions in the same way also have the same function. They produce a classification that resembles the ones described before (Maurus, Kühlmorgen & Hartmann, 1973).

(e) Variability in duration of communication processes

For classification purposes, frequency changes in the occurrence of units may be considered even if the changes occur from one communication process to the next. Such changes, notably in the duration of a process (expressed by the number of events), were constantly observed in our

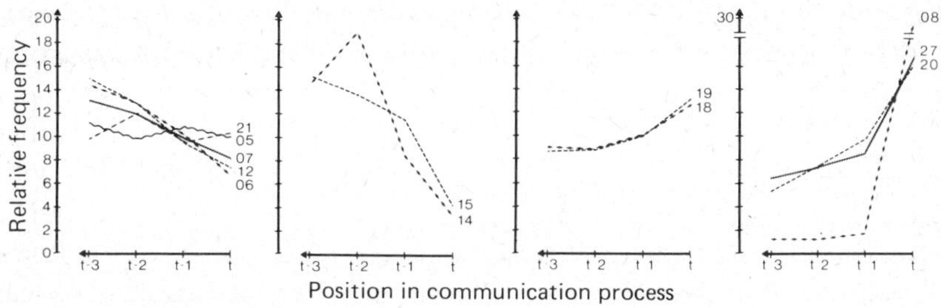

Figure 11.2. Course of the frequency of several behaviour units over the final four positions (t−3, . . . t; abscissa) in a communication process. For a certain unit and a certain position in the sequence the frequency of this unit at this position is divided by the number of times that any unit occurs at this position and by the frequency of this unit in the experiments. The number labelling each curve corresponds to the number of the behavioural unit (see Table 11.1).

experiments. When assuming that the duration of a communication process correlates somehow with the functional content of a signal (e.g. in agonistic processes of extended duration, the signals exchanged may be of higher intensity than those in briefer processes), a classification that takes account of functional aspects may be derived: of all sequences having the same duration (e.g. three events), the total number of occurrences of each unit can be counted and then compared with the frequencies with which these units occur in longer sequences (e.g. eight events). This will result in a categorization which again resembles the classification results by the other methods (Maurus *et al.* 1979).

There are still other approaches to classification which cannot all be discussed here. Let it suffice to say that all of them confirm the classification results achieved so far. This agreement among the results, derived, as they are, from various independent approaches to the assessment of the available data and calculated by applying various mathematical models, leads us to assume the validity of the results obtained. An additional verification of our classification may be found in the following considerations.

When assuming that intra-specific communication processes are predetermined to a certain extent (which can be derived from the definition of the function of a signal, i.e. its effect on the recipient), the description of communication processes based on functionally defined units is bound to produce a greater measure of predictability than a description disregarding the function of units.

A measure of the predictability of the succeeding event, the preceding event being known, is provided by the quantity 'transinformation'. It can be shown that this quantity assumes the most favourable value if the communication processes observed are described in terms of the classes established (Pruscha & Maurus 1973).

Interpreting the function of classes

In order to recognize the function of a signal, it is necessary to know the function of its effect. Usually, this effect manifests itself in a response pattern which is a signal in itself. For the adequate interpretation of such evoked signals, previous knowledge of the functions of at least some of these signals is important. So, if the sender can be induced to give signals that cause recipient reactions of known function, the function of signals as yet unknown could be interpreted from the recipient's reactions. However, when the function of none of the observed signals is known, or the sender cannot be induced to give signals of unknown function that cause recipient reactions of known function, then the direct approach to analysing the function of signals is blocked. What is to be found are indirect approaches to function analysis.

While the classification methods based on analyses of transition frequencies between preceding and succeeding behavioural events provide indications as to whether two different units are similar or dissimilar in function but not as to the type of function, other classification methods furnish additional data from interpreting the function of the classes established.

Where the classification is based on the intensity of the effect an event has on the succeeding events, all units with the same intensity may be assigned the same function. Differences in intensity are an indication that the units concerned must not be assigned to the same category. Furthermore, to determine the degree of intensity of a signal serves also to determine which of the classes distinguished has the higher intensity of effect on the succeeding behaviour, and which has the lower. In the field of agonistic behaviour it may be assumed that signals with higher effect intensity are those that express aggressiveness of higher intensity. This assumption results in an order of classes reflecting the aggressive effect of their units.

This order can also be derived from the positioning of the categories in the communication processes and from the frequency of their occurrence in communication processes of various durations.

Agonistic communication processes evoked by electrical brain stimulation are characterized by the fact that, in a situation where no discernible agonistic events are expected, the aggressive behaviour of one animal may all of a sudden cause these agonistic interactions to begin. After such encounters, peaceful behaviour will soon be resumed. Classes that often occur at the end of such encounters may be assigned a lower intensity than classes rarely or never observed at the end of encounters.

Besides this intensity scale of aggressive behaviour, there are other classes that do not fit into this linear system. A characterization of these classes may also be derived from the properties they show in communication processes. It is difficult, however, to label them with the usual terms. An attempt at solving this problem is shown in Table 11.1. The list in that table does not claim to be exhaustive, neither in terms of the repertoire in the field of visually recognizable behaviour, nor in terms of the repertoire of other sensory modalities. The inclusion in our analyses of other parts of the overall repertoire, especially acoustic signals, is currently being prepared.

The units listed in Table 11.1 have been arbitrarily selected with a view towards meeting the requirements of the classification method. The categories and their functional order are the result of the quantitative methods described. They have been labelled arbitrarily. The names they have been given may be contested – depending on how important a dispute over these terms is considered to be.

Because of the incompleteness of the repertoire, our results should be considered as temporary and probably not differentiated enough. However, we are confident that an increase in the repertoire will provide more categories and allow for better differentiation. On the other hand, the temporary restriction to a narrowly defined sector of the behavioural repertoire should not be interpreted as mainly negative. The fact that the multi-directional method presented in this paper has led to consistent results, even if the available repertoire of an animal is not complete, is a definite asset for the application of this method. In addition to offering the chance of starting successful analyses even when the repertoire is still incomplete, there is the advantage that errors attributable to the arbitrary and intuitive determination of behavioural units can be identified and corrected. Furthermore, it has been shown that these analyses can be performed successfully even when electrical brain stimulation cannot be applied (Maurus *et al.* 1981).

Our categorization results have probably not contributed much to answering the question of the meaning of primate signals as defined at the beginning of this paper. Neither have those results opened any approaches to quantitative analyses in this respect. Still, we believe that a categorization method that proves its validity by providing a degree of predictability that is better than that of all other repertoire classifications will be a useful basis for seeking an answer to the question of the meaning of primate signals.

11.3. Special problems in classifying acoustical signals

Communication processes by squirrel monkeys consisting exclusively of visually recognizable signals are extremely rare. Most of these processes consist of a series of both visually recognizable and acoustical signals. It is quite possible that signals from differing modalities mutually influence each other's function, or that one signal consists of several components from differing modalities. Strictly speaking, the analysis of signal functions may be considered truly comprehensive only if all modalities are taken into account. An analytic approach that, right from the outset, were to take such a highly complex structure into consideration could hardly be expected to be very successful. A more promising approach would be to start with simple steps and then gradually increase their complexity. This means analysing acoustical signals first without regard to the visually recognizable signals to be considered later.

It is possible to use intuition when defining the units, where visually recognizable signals are involved: when watching the animals, it is comparatively easy to distinguish the various movements and positions. Defining acoustical units is far more difficult, as evidenced by the reports of several authors (Winter, Ploog & Latta 1966; Schott 1975; Jürgens 1979). They all defined units on the basis of both acoustic impressions and sonograms, but each of them came up with a somewhat different classification. This is an indication that – as in the case of visual signals – the observer is not always capable of distinguishing acoustical signals, while the animals clearly are, or that the observer may notice a difference where there is none as far as the animals are concerned.

Seyfarth, who conducted experiments with vervet monkeys, mentions one example (this volume) where the observer, despite listening to vocalizations and comparing sonograms, was unable to perceive in his first approach any difference in the signals which might be essential for distinguishing the various functions. That various functions had, in fact, been assigned to supposedly one and the same call was proved by Seyfarth, by analysing the receiving animals' reactions to it.

Such safely reproducible reactions have not been registered so far in our experiments with the squirrel monkey, neither by loudspeaker playback of calls nor by eliciting vocalizations by telestimulation. For the time being, this rules out the functional classification of the vocal repertoire on the basis of the receiving animals' reactions.

What remains is an approach via a classification method similar to the one successfully applied in classifying visually recognizable signals – provided, of course, we know which of the calls to define as units in order to make them part of a classification method. Finding these units under conditions dominated by the squirrel monkey would be the next step towards the incorporation of acoustical stimuli in the function analysis of signals.

Postscript on language and brain structure

We wish finally to bring home the major question of whether linguistic concepts and methods of analysis are adequate means for the description and explanation of the communication of non-human primates. In biology and other fields of the natural sciences there is a general rule by which the causation of events at more complex levels of organization is explained by processes at less complex levels of organization: e.g. certain events in chemistry – say forces in a molecule – are explained in terms of physics and certain events in physiology are explained in terms of physics and/or chemistry. The reverse procedure, i.e. the explanation of physical events in chemical or physiological terms, would not make sense.

In the case of primate communication we all agree that it is not human language that is used. Since human language is certainly more complex than primate communication, I doubt that concepts and rules derived from linguistics are applicable in the analysis of communication processes in primates. This conjecture is supported by our neuroanatomical and neurophysiological findings concerning primate vocalization which is a substantial part of the animal's communication. Occasionally, students of this problem use the term 'the organ of language', referring to the brain mechanism which produces language. We know from neuropsychology and aphasiology that the cerebral cortex plays an essential role in the expression of language. In the monkey it does not; the complete vocal repertoire can be produced without the neocortex. There is, however, as we have shown, a whole system in the monkey's brain, which is necessary for the production of vocalization. This phonatory organ reaches from subcortical, limbic structures of the brain down to the midbrain and medulla. Putting this in an evolutionary perspective, one can say that the phonatory organ of the brain was present long before the language organ evolved, the former is at a less complex level of organization than the latter. There is yet another level which one might call the 'articulatory organ of the brain', which has its separate function. All three 'organs' are necessary for producing human speech.

References

Jürgens, U. (1979). Vocalization as an emotional indicator, a neuroethological study in the squirrel monkey. *Behaviour*, **69**, 88–117.

Kummer, H., Goetz, W. & Angst, W. (1970). Cross-species modifications of social behavior in baboons. In *Old World Monkeys*, ed. J. R. Napier & P. H. Napier, pp. 353–63. New York & London: Academic Press.

Maurus, M. (1967). A new telestimulation technique for the study of social behavior of the squirrel monkey. In *First Congress of International Primatology Society, Frankfurt*, ed. D. Starck, R. Schneider & H.-J. Kuhn. Stuttgart: Fischer-Verlag.

Maurus, M., Geissler, B., Kühlmorgen, B. & Wiesner, E. (1981). The effects of brain stimulation when categorizing the behavioral repertoire of squirrel monkeys. *Behavioral and Neural Biology*, **32**, 438–47.

Maurus, M., Kühlmorgen, B. & Hartmann, E. (1973). Concerning the influence of experimental conditions on social interactions initiated by telestimulation in squirrel monkey groups. *Brain Research*, **64**, 271–80.

Maurus, M. & Ploog, D. (1971). Social signals in squirrel monkeys: analysis by cerebral radio stimulation. *Experimental Brain Research*, **12**, 171–83.

Maurus, M. & Pruscha, H. (1972). Quantitative analyses of behavioural sequences elicited by automated telestimulation in squirrel monkeys. *Experimental Brain Research*, **14**, 372–94.

Maurus, M. & Pruscha, H. (1973). Classification of social signals in squirrel monkeys by means of cluster analysis. *Behaviour*, **47**, 106–28.

Maurus, M., Pruscha, H., Wiesner, E. & Geissler, B. (1979). Categorization of behavioural repertoire with respect to communicative meaning of social signals. *Zeitschrift für Tierpsychologie*, **51**, 48–57.

Ploog, D. (1979). Phonation, emotion, cognition, with reference to the brain mechanisms involved. In *Brain and Mind*, Ciba Foundations, Series 69, pp. 79–98 Amsterdam: Excerpta Medica.

Ploog, D. (1981). Neurobiology of primate audio-vocal behavior. *Brain Research Review*, **3**, 35–61.

Pruscha, H. & Maurus, M. (1973). A statistical method for the classification of behavior units occurring in primate communication. *Behavioural Biology* **9**, 511–16.

Pruscha, H. & Maurus, M. (1979). Analysis of primate communication by means of a multiresponse linear learning model. *Revue Roumaine de Mathématiques Pures et Appliquées*, **24**, 1371–83.

Reynolds, V. (1976). The origins of a behavioural vocabulary: the case of the rhesus monkey. *Journal for the Theory of Social Behaviour*, **6**, 105–42.

Schott, D. (1975). Quantitative analysis of the vocal repertoire of squirrel monkeys (*Saimiri sciureus*). *Zeitschrift für Tierpsychologie*, **38**, 225–50.

Winter, P., Ploog, D. & Latta, J. (1966). Vocal repertoire of the squirrel monkey (*Saimiri sciureus*), its analysis and significance. *Experimental Brain Research*, **1**, 359–84.

COMMENT

ROM HARRÉ

Once the propriety of intensional modes of description has been admitted, an important consequence follows: in order to carry out an analysis of a stream of activity to reveal its structure, element-types must be differentiated to classify and identify units of action. But in an intensional mode of description these are conceptual or meaning units, *not* behaviour units. So different behaviour units may be realizations of the same meaning unit, and so collected under the same concept. In practice, this means that semantic hypotheses are needed in order to make a formal syntactical analysis possible. But the meanings of actions are themselves defined relative to the presumed social process of which they are the performance. It follows that a third level of hypotheses is also implicated, namely social hypotheses. The conceptual foundations of behaviour analysis are therefore hierarchical.

Social hypotheses (perlocutionary force)
↓
Semantic hypotheses (illocutionary force)
↓
Syntactical hypotheses

It follows immediately that a study of syntactical structure cannot be used to establish semantic hypotheses, i.e. to establish criteria of demarcation and partition of actions inductively, since it presupposes these hypotheses.

The technique of Maurus & Ploog is not a way of discovering semantic units, one by one. Rather the technique is a way of using a prior (implicit) system of hypotheses about significant partitionings of activity to discover behaviour patterns. The test for whether the results are adequate can only be by assessing the plausibility of the whole segment of the total life form, as the end product. The technique cannot pick out meaningful actions one by one.

This point can be neatly illustrated in the following example. Suppose we identify four action units, A, F, C and B (as in the authors' illustration), and we want (as do Maurus & Ploog) to test the hypothesis that A and F are distinct signals, that is are semantically distinct. Using these preliminaries, we record the following empirical patterns:

A is followed by C in 10% of cases and by B(i) in 90% of cases.
F is followed by C in 10% of cases and by B(ii) in 90% of cases.

These results refute the hypothesis that A and F are semantically identical only on the hypothesis that a B following an A (B(i)) is semantically identical with a B following an F (B(ii)). There can be no morphological proof of that hypothesis, since the physical shape of the action and its meaning can never be presumed to be in 1:1 correspondence. The correct conclusion to draw from the results above is the conditional one, that A and F are not semantically distinct, *if* B(i) and B(ii) are semantically identical. Relative to the above results, the presumed identity 'B(i) = B(ii)' is a hypothesis. We could test this hypothesis only by relating B(i) and B(ii), that is 'B after A' and 'B after F' to further and different features of the social life of the creatures involved, and so on. In the experiments, each particular hypothesis, like the above about A and F, is tested relative to its relations to the totality of a form of life.

Again admitting intensional descriptions introduces an element of the hypothetical into the interpretation of fixed adjacency pairs, that is pairs of actions whose transition probabilities are always 1. Such pairs might appear because in a form of life they constitute one action unit (kissing first one cheek then the other is *one* greeting-act but *two* kisses, with transition probability 1). But it might be because they are distinct action units, the second obligatory or required on the occurrence of the first. Thus 'hold tail/bite tail' may be found with transition probability 1 because the monkey holds to the belief that 'if holding tail, must bite tail'. (Compare the following: 'If plural subject, must pluralize the verb', 'If thanked ("Thanks") must issue disclaimer ("You're welcome")'.) We can only know whether 'hold tail/bite tail' is one action or two actions by reference to the rest of the form of life. Whichever we take it to be will be a conclusion conditional on our background assumptions.

REPLY

M. MAURUS AND D. PLOOG

Harré's statement that, in order to prove the semantic identity of action units A and F, action unit B following action unit A and action unit B following F must be semantically identical, is absolutely correct. And when he states that a 1:1 correspondence cannot be presumed for the semantic identity of a unit defined by intuitively interpreting positions (physical shape) and reactions, that is also correct. That is why we mentioned in our paper that a basic requirement for our work was that none of the units selected had more than one function. When observing this requirement, B following A and B following F will have the same function (semantic identity). However, for ascertaining that each unit is actually assigned only one function, to make this requirement will be of little help if the observer does not known how to establish it. So he will start by dividing the behavioural repertoire into relatively many units. This way he can be sure that at least some of the units registered will meet the requirement. Next, he will check each unit against all the others, i.e. he will treat each unit as a preceding as well as a succeeding signal (see Figure 11.1). The analyses will reveal whether B has more than one function. If so, B will be further divided and the analyses repeated with the newly defined units.

In dividing the units, all aspects assumed to be of functional importance should be considered, such as variations in motor patterns, in positions as well as possible functional changes due to contextual dependence. By the analyses we developed (e.g. Maurus & Pruscha 1972; Pruscha & Maurus 1976), we were able to prove such functional dependence. When a unit has been found to be dependent on its context, it will serve, together with the findings of the analyses, as a basis for identifying new units which will have only one function. In conducting these analyses, the total amount of units on hand will be considered, not only the ones selected for further division. Upon conclusion of the analyses, (that is, after each unit has been checked against all the others), those units (or classes) with only one function will remain. This, of course, applies only to those parts of the behavioural repertoire to which we had to restrict ourselves. For instance, our analyses covered only visually recognizable units; they disregarded the influence acoustical signals may have.

Now for the other point mentioned by Harré: the transition sequences A→C, A→B, F→C and F→B are supposed to serve as examples of the sequence signal→discernible effect. Precise 1:1 transitions observed in this connection would be absolutely exceptional – as far as the squirrel monkey is concerned – and would be an ideal signal→discernible effect situation making further, more complex analyses unnecessary. When such a 1:1 transition is successively observed in one and the same animal (if I interpret the examples Harré gave correctly), it is not a sequence of signal and discernible effect but a signal consisting of two units with one always following the other. These two units could then be combined into one unit and included in the signal function analyses – of course, only after having made sure that neither one of them ever appears alone.

References

Maurus, M. & Pruscha, H. (1972). Quantitative analyses of behavioural sequences elicited by automated telestimulation in squirrel monkeys. *Experimental Brain Research*, **14**, 372–94.

Pruscha, H. & Maurus, M. (1976). The communicative function of some agonistic behavior patterns in squirrel monkeys: the relevance of social context. *Behavioral Ecology and Sociobiology*, **1**, 185–214.

12. Experience tells

ERIC JONES AND MICHAEL CHANCE

Individuals of social species of monkeys and apes cohere in groups, that is to say, each individual has a built-in propensity,[1] expressed in diverse ways of behaving, that enables it to remain a member of the group.

Primatologists have been slow to give this the priority it merits for understanding the way social life is the single most urgent factor in the organization of the individual's behaviour. Kaplan (1978) in his study of *Macaca mulatta* showed that half the threats given by one monkey in support of another could be ascribed to the individual's attempt to become closely associated with individuals of higher status (the other half being in support of kin). That this socializing tendency is expressed in a centric organization of social attention has been demonstrated by Emory (1975). He has also shown that it is greater in an open savannah-living species (*Therophithecus gelada*) than in a riverine forest species (*Mandrillus sphinx*). Hence, this built-in propensity sets a purpose for the maturing individual, and is evident in the herding behaviour of dominant males, for example in *Papio hamadryas*. Our studies (Chance & Jones unpublished) of *Macaca fascicularis* strongly suggest that the monitoring of the social behaviour of individuals in his group is a constant concern for the dominant male. He restrains excessively aggressive females; this behaviour in captive colonies is associated with damage to the aggressee. In feral groups, aggression is likely to lead to the disruption of the group by dispersion. The evidence is, therefore, that a critical function for the selection of a socializing tendency is security through membership of the group. This is based on cohesion around the dominant individual from whom protection is obtained, not only against predators but – for females – against disruptive aggression by other females. This concern is first seen in the protection of the infant by the mother. It is not surprising, therefore, in our long-term study of our *Macaca fascicularis* group in captivity we have found diverse social strategies to be an integral part of the social behaviour.

It is unlikely that mature socially adjusted monkeys will make many mistakes in their interpretation of signals arising from the environment, particularly in signals originating in the behaviour of other group members. The reason for this is clear: they have learned what behaviour to expect from the other members of the group.

All monkeys are held in the nexus of the group's social structure, each with various limitations on the freedom of its behaviour, imposed upon it by its basic rank, its dependent rank – ranking based on the presence of supporters, allies, or what we term referents. Individuals to whom another pays significantly more attention than to others are described by us as social referents. Negative social referents receive a high proportion of another's gaze and are continuously avoided. Positive referents receive a significantly low amount of gaze from another and a large amount of neighbourly behaviour. A positive referent may act as a backer by virtue of being of higher status and allowing the referee to threaten from nearby, or may actively support the referee by joining in threats against a third individual (Pitcairn 1976). Other 'factors' obviously interact, particularly the influence of maturation, state of sexual cycle in females, time of year (breeding seasons, etc.).

We suggest that there are two areas where one could expect mistakes to be made in an animal's interpretation of signals from a conspecific. Firstly in early infancy the infant has to learn how to relate to others (initially to its mother), then increasingly to other group members. It is here, therefore, that we shall present some examples of failure to interpret the signals emanating from the environment.

Secondly, misunderstandings between animals are likely to occur where extreme social changes have taken place. We have documented a rebellion within our colony of monkeys (Chance, Emory & Payne 1977). In the post-rebellion period we found that the animals obviously fully comprehended the signals that were being emitted by their companions. What confused them was that the social situation had changed.

To return to the mother/infant relationship; we can consider a monkey mother and her infant as two cog-wheels. The infant makes demands on the mother but also, to a lesser extent, the mother makes demands on the infant. When the demands are synchronous there is a meshing of the cogs and the relationship is smooth. However, when the demands do not mesh – a mis-mesh – the relationship is not smooth, it grates. To obtain mesh the mother has to be able to 'read' the infant's signals and correspondingly the infant has to be able to read mother's signals. Naturally, given the helplessness of the new-born in early life, it will be largely the mother's responsiveness which will determine whether or not there is meshing in the relationship.

We have shown (Chance, Jones & Shostak 1977) that nursing mother monkeys show meshing in terms of the time they spend nursing together – nursing involves a positive influence, or imitation of one nursing pair by the other, and is imitated by other members of the group who nurse each other. This ensures a coincidence of mood and hence the likelihood of meshing. Nursing, nevertheless, provides an area where the most mis-mesh in the mother–infant relationship can occur. This change begins to appear very early in this relationship. We have shown that monkey mothers begin to withdraw their nipples from the infants' mouths as early as 10 days, and to prevent access to the nipple a few days later. One method, used when the infants were very young, was simply to lift up the whole shoulder region on the nipple side. An extension of this method used with older infants was to stretch the arm up as the shoulder was lifted, by these methods the nipple could be pulled out of the infant's mouth. Another, apparently more gentle, method of nipple withdrawal was when the mother slid a finger down the breast and into the infant's mouth, thus substituting the finger while the nipple was being extracted. This method of nipple removal did not seem to distress the infants as much as some other techniques.

Prevention of access to nipple was again brought about by several methods. With younger infants the nipple was lifted 'out of range' of its mouth by techniques similar to nipple withdrawal. A simpler method was to interpose an arm between the infant's body and the breast. Persistent attempts to gain nipple contact were often 'punished' by the method previously mentioned, 'punitive deterrent'. When the infant was older and was obviously approaching the mother with a view to gaining the nipple (this was readily determined by the human observer, for under these circumstances the infant's visual attention was focused on the nipple) the mothers would often simply move away. In terms of withdrawal-prevention, a peak is achieved at 4–5 weeks. The response of the infant to these nipple-withdrawal rejections is the so-called weaning tantrum. These are first seen at around one month of age (the time of peak rejection), and become very infrequent after about three months. They take the form of 'geckering' bouts, sometimes accompanied by extremely erratic behaviour.

We have seen some 171 so-called weaning trantrums, and have been able to show that some of the more intense forms (hitherto called 'convulsions', 'fits' and 'spasms' seen by other workers) are really extended startle reactions (Chance & Jones 1974). We have been able to induce these experimentally and have come to the conclusion that they may result from a difference between the mother and the infant in the interpretation of the same signal. Take the case of a threatening stimulus, say an unfamiliar human outside the cage, or a dog visible through the window. Up to the age of about 12 weeks the mother will run to the infant and retrieve it, the infant also escaping from the stimulus to the mother. Here is a mesh between the mother's behaviour and the infant's, i.e. they interpret (understand) the stimulus in the same way. After the age of about 12 weeks the infant flees to the environmental refuge, i.e. upwards, and only to mother in special circumstances.

Similarly, take the case of a younger infant which finds an environmental stimulus startling. Its response is to flee to mother. If, however, the stimulus is one with which the mother is familiar, she will not be startled – she may, indeed, be unwilling to take up the infant. We now have a mis-mesh between the reactions/moods of the dyad. We have indicated that mere rejection from nipple contact may cause 'tantrums', but here we have a situation where an already aroused infant is seeking comfort and its 'comforter' (mother) rejects it. This augments the arousal and so the infant goes into a 'protracted startle' – now fleeing from mother. The movement away from mother when she should be the refuge, is in itself, more arousing so the infant may enter an equilibratory phase, rapidly oscillating between departing and returning to mother, giving distress vocalizations. A full report of this behaviour has been given (Chance & Jones 1974).

We have concentrated on the mother/infant relationship. We feel that in the search for 'understandings' between monkeys, a major source of information may lie in the study of this relationship. Indeed, a study of inexperienced (primiparous) mothers may reveal many 'misunderstandings' due to different interpretations of the same signal. It may, perhaps, be said that one of the most important aspects of learning in infancy is learning to interpret social signals of conspecifics and learning the limits of one's freedom of behaviour.

Notes

1. Propensity: latent form of behaviour which needs practice to bring into operation.

References

Chance, M. R. A., Emory, R. G. & Payne, R. G. (1977). Status referents in long-tailed Macaques (*Macaca fascicularis*): Precursors and effects of a female rebellion. *Primates*, **18**, 611–32.

Chance, M. R. A. & Jones E. (1974). A protracted startle response to maternal rejection in infants of *Macaca fascicularis*. *Folia Primatologica*, **22**, 218–36.

Chance, M. R. A., Jones, E. & Shostak, S., (1974). Factors influencing nursing in *Macaca fascicularis*. *Folia Primatologica*, **27**, 28–40.
Emory, G. R. (1975). The patterns of interaction between the young males and group members in captive groups of *Mandrillus sphinx* and *Theropithecus gelada*. *Primates* **16**, 317–34.
Kaplan, J. R. (1978). Fight interference and altruism in Rhesus Monkeys. *American Journal of Physical Anthropology*, **49**, 241–50.
Pitcairn, T. K. (1976). Attention and social structure in *Macaca fascicularis*. In *Social Structure of Attention*, ed. M. R. A. Chance & R. R. Larsen. London: Wiley.

COMMENT

DUANE QUIATT

Jones and Chance's reference to the unlikelihood of normal adult monkeys making many mistakes in the interpretation of signals, particularly signals which carry information about the behavior of other group members, underlines the importance of socialization and continual information exchange to recognition of signal specificity, relative not only to other signals in a species inventory but to equivalent signals emitted by different individuals.

What about the same individual sending equivalent signals under slightly different circumstances? Cheney notes that if signals (e.g. alarm cries, grunts) were no more than responses to general arousal they would not carry much information. Can we talk about signals as indicative of more specific arousal states? Is this useful terminology, and what would it mean to use it?

Dawkins & Krebs (1978), in an illuminating but one-sided analysis, see animal communication in terms of actors who use signals to manipulate reactors, increasing their own energetic efficiency through the use of others' muscle power. In Dawkins & Krebs' view, one hardly need refer to the informational content of signals. But the converse holds too (see my note 3, p. 37). Signal recipients can, in effect, rely on other organisms as detached environmental sensors and first order data processing systems. If it makes sense to talk about signal reception in these terms, then it seems to me we can make useful comparisons of information content even if we want to think of signals as no more than indicators of arousal states, and of communication between group associates (for instance) as no more than an exchange of readings of internal environmental states.

Socially adjusted adult monkeys may rarely misunderstand one another, but, as Jones and Chance note, inexperienced mothers may frequently misinterpret the signals of their offspring. While the function of allomaternal behavior has been hotly debated for some years now, the proponents of the 'learning-to-mother' hypothesis have never been very explicit about just what it is that is learned. It seems clear that handling a number of different infants would provide excellent training in learning to 'read' signals that were functionally equivalent despite individual idiosyncracies in transmission. Experience thus gained would seem to provide a comparative framework for efficient subsequent sorting of one's own infant's variable signals into reliable categories. It may be that that sort of learning could be accomplished with less emotional distortion than if one had only one's own offspring from which to learn.

References

Dawkins, R. & Krebs, J. R. (1978). Animal signals: information or manipulation? In *Behavioural Ecology, an Evolutionary Approach,* ed. J. R. Krebs & N. B. Davies. Oxford: Blackwell.

REPLY

ERIC JONES AND MICHAEL CHANCE

Quiatt's comments are well made. We would only point out that he perhaps overrates the mother monkey's role, for the infant monkey has also to learn how to interpret its mother's signals in order for the relationship to 'mesh' smoothly.

Prospects for future research

Future research will have to tackle two problems: the structure of primate mentation processes themselves, and how to describe them. The contributions to this book suggest several topics for further study under those two broad heads. As regards the former, a question that is certainly worthy of more thought is the extent to which the cognitive processes of monkeys can be considered to be in some ways structured and limited by evolved propensities. Whether or not human thought is capable of 'running free' of innate programmes, the same is unlikely to be true of primate mentation.

Further work on the kinds of discriminations incorporated into primate signals will again almost certainly be done, doubtless using refinements of the Seyfarth–Cheney playback technique. Perhaps this could be done with visual or olfactory or even tactile signals. In the social sphere, the field of kinship suggests itself as an area in which the extent of social discrimination could be explored.

If the field of teaching sign language to apes is now a somewhat overworked area, we still do not know quite what all those famous chimpanzees and gorillas achieved. We need to know how close their achievements come to the use of language and the underlying conceptual processing found in humans. Part of the problem is that the abilities of chimpanzee *outside* of sign language are still imperfectly understood. We need to know much more about the kinds of discrimination made by wild or free-ranging chimpanzees in their normal, untutored everyday lives. The same problem hit animal psychology years ago, with the advent of ethology. Suddenly the laboratory rat was shown to have, not a blank brain on which the experimenter wrote his instructions, but a brain already containing a huge and complex set of interlocking programs. Chimpanzee too have inborn tendencies and their performances in learning situations have to be seen in terms of these.

The second major problem area is that of description and interpretation. How should we proceed? There are two interlocked issues: how to justify the use of intentional language for describing some aspects of primate behaviour; and how to make a start on resolving the problem of whether an intensional account of animal thought is appropriate.

The first of these issues is liable to be muddled with the long-standing debate about the propriety and limits of anthropomorphism.The question of the quality of animal experience is a quite separate puzzle from the problem of finding the best vocabulary to describe animal life. To speak of chimps as pondering, prevaricating, remembering, anticipating, planning, reflecting and so on can be justified by a research programme in which particular incidents (anecdotes too) are collected and contrived as 'test beds' for vocabulary. Would any expression other than an item from the intentional vocabulary have picked out just exactly *this* event in all its particularity and complexity? We cannot see that at the present time there is any interest in debating the nature of the quality of the experiences of primates. So the question of whether primates are conscious, as a phenomenological query, seems otiose. The fact that these vocabulary-testing situations are public events does not imply a behaviourist reduction of animal psychology. It no more follows that an ape's pondering is just its sitting apart with a far away look than the use of a similar look to pick out human reflection and ratiocination implies that human pondering is just sitting and looking a certain way.

The other problem – whether an intensional account of such animal cognition as planning or expecting is required – has to be tackled another way. Conceptual thought is abstract whereas iconic representations (imaginings, say) are concrete. We can reasonably treat cognition as intensional where the task to be undertaken on the basis of thinking is of a certain order of sophistication. None of the instances cited in the work reported in this book, we believe, makes an unequivocal demand for an intensional treatment of the thought that went into carrying it out. Some primate 'language' studies have been directed to pushing up the level of sophistication of the 'conversation' between man and ape, so that the ape may be seen to be manipulating abstractions. But by and large whatever has been found to hint at a higher level of abstraction in thought has turned up serendipitously in the course of studies directed to other ends. A systematic attempt to push the tasks confronting primates to higher levels of abstraction might lead to some resolution of the presence of modes of representation.

Name index

Subject index